Methods for Analysis of Carbohydrate Metabolism in Photosynthetic Organisms

Methods for Analysis of Carbohydrate Metabolism in Photosynthetic Organisms

Plants, Green Algae, and Cyanobacteria

Horacio G. Pontis

Professor Emeritus of Biochemistry, Universidad Nacional de Mar del Plata, and Emeritus Senior Investigator and Vice-President, Applied Biological Research Foundation — FIBA, Buenos Aires, Argentina

AMSTERDAM • BOSTON • HEIDELBERG • LONDON
NEW YORK • OXFORD • PARIS • SAN DIEGO
SAN FRANCISCO • SINGAPORE • SYDNEY • TOKYO
Academic Press is an imprint of Elsevier

Academic Press is an imprint of Elsevier
125 London Wall, London EC2Y 5AS, United Kingdom
525 B Street, Suite 1800, San Diego, CA 92101-4495, United States
50 Hampshire Street, 5th Floor, Cambridge, MA 02139, United States
The Boulevard, Langford Lane, Kidlington, Oxford OX5 1GB, United Kingdom

Cover design: Sheila Pontis

British Library Cataloguing-in-Publication Data
A catalogue record for this book is available from the British Library

Library of Congress Cataloging-in-Publication Data
A catalog record for this book is available from the Library of Congress

ISBN: 978-0-12-803396-8

For Information on all Academic Press publications
visit our website at https://www.elsevier.com

ELSEVIER
Book Aid
International
Working together
to grow libraries in
developing countries
www.elsevier.com • www.bookaid.org

Publisher: Sara Tenney
Editorial Project Manager: Mary Preap and Joslyn T. Chaiprasert-Paguio
Production Project Manager: Lucía Pérez
Designer: Alan Studholme

Typeset by MPS Limited, Chennai, India

To Samantha, who accompanied me on my journey writing this book

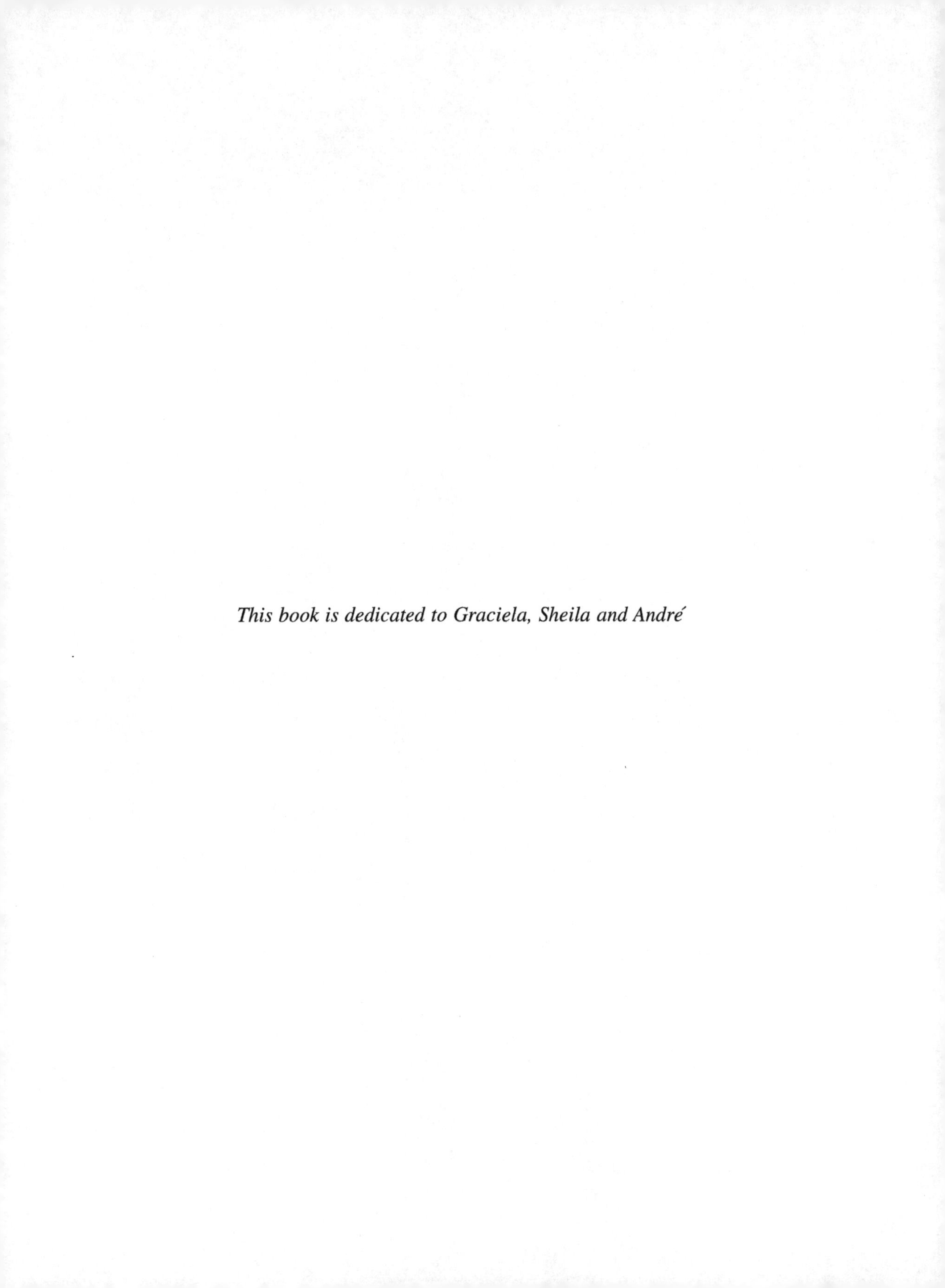

This book is dedicated to Graciela, Sheila and André

This book is dedicated to Gretchen, Sheila and Amelie

Contents

PART III: CASE STUDIES

Biography

Horacio Guillermo Pontis was born in Mendoza, Argentina in 1928. He graduated from the University of Buenos Aires chemistry school and obtained his doctorate working in organic chemistry in 1951. He worked alongside Professor Luis F. Leloir for three years, who influenced his interest in carbohydrate metabolism, sugar phosphates, and sugar nucleotides. He spent a long stay at King's College, Durham University (United Kingdom) and at Karolinska Institutet, University of Stockholm (Sweden), where he started enzymology studies with Prof. Peter Reichard. After his return to Argentina in 1960, he began his work on the field of plant biochemistry, studying fructan and sucrose metabolism. In connection with these studies, he synthesized fructose-2-phosphate, the first ketose that allowed other researchers the chemical synthesis of fructose-2,6, diphosphate (a key glycolysis activator) two decades later. Since 1961, he has been a member of the National Research Council of Argentina and Professor of Biochemistry, firstly at University of Buenos Aires and then at University of Mar del Plata. Between 1967 and 1977, Pontis was Director of the Biology Department of Fundación Bariloche, and also in 1973, he was President of the Sociedad Argentina de Investigación Bioquímica (SAIB). In Mar del Plata (Argentina), he set up the Institute for Biological Research at the University, and together with Prof. Leloir, they established the Foundation for Biochemical Applied Research (FIBA), where as Head of its Biological Research Center, he maintained an active research group, training graduate and postgraduate students, and producing a steady flow of research publications.

At present, Pontis is Professor Emeritus of University of Mar del Plata, Emeritus Researcher and vice-president of FIBA. His important contribution to plant biochemistry was recognized by the Argentinean Society of Plant Physiology and by the American Society of Plant Biologists, which awarded him as a corresponding member.

Preface

The idea of writing a book on the analytical methods for the study of carbohydrates, essential components of the central metabolism of land plants, unicellular green algae, and cyanobacteria, came to me at the insistent request of my best student who insisted that I should pass on the experience I had gained over so many years devoted to this field.

At first, the request seemed simple to implement; however, I had to tell my experience in a way that no other published texts had done so before, where general principles of separation and purification of proteins or methods to quantify a given carbohydrate were provided. It suddenly occurred to me that the best way to achieve this goal was to describe the different methodologies that I had been extensively using in my laboratory as if I were orally explaining them to my students or colleagues who were not familiar with them. My intention was, to some extent, to show that carbohydrates and their metabolic pathways could be studied by following quick and easily accessible methodologies.

Building on my idea, I decided to describe in detail each experiment to be performed, starting from the technique's fundamentals or principles, mentioning the necessary steps for implementation, also including the biological starting material, and the choice of methodology among different alternatives. To put it into a few words, I intended to provide researchers with tools and information ranging from an overview of the extraction and purification of the enzymes involved in carbohydrate metabolism from photosynthetic organisms to different analytical techniques for the measurement of their activities, and separation and determination of sugars and other compounds related to them. To me, the clearest way to transmit this information was to apply it to the study of specific cases, selecting the key sugar players. I trust readers will find this approach useful and that it will assist those who are taking their first steps into the field of sugar metabolism.

Perhaps the most important aspect of this book, and what distinguishes it from the literature available, is that it brings to the present a selection of reliable and tested methods used decades ago but not widely applied these days, which do not require costly instruments and specially trained personnel, together with some more modern techniques.

Regarding the book organization, it has been divided into three parts for an orderly and systematic description of the general analytical methods. Part I (see Chapter 1: Determination of Carbohydrates Metabolism Molecules) focuses on procedures for determining the most relevant carbohydrates and some compounds related to their metabolism, including colorimetric, spectrophotometric, and spectrofluorimetric methods. The applications of these methods are presented in later chapters. Part II starts with the description of general methods for the extraction of proteins with enzymatic activity involved in sugar metabolism (see Chapter 2: Preparation of Protein Extracts). Chapter 3, Protein and Carbohydrate Separation and Purification describes the most common methods for separating proteins (ion exchange, gel filtration, affinity chromatographic methods, isoelectrofocusing, and fast protein liquid chromatography) and carbohydrate fractionation (chromatographic methods, including HPLC (high-performance liquid chromatography) and HPLC coupled to mass spectrometry). The determinations of enzyme activities, measured as either substrate consumption or as product appearance, are reported in Chapter 4, Measurement of Enzyme Activity. The next chapter deals with a general overview of the fundamentals of the methodologies based on mass spectrometry (MS) and nuclear magnetic resonance (NMR), which despite not being laboratory bench methodologies are sometimes resorted to through external services. Specific cases are covered in Part III, starting with sucrose, the main plant sugar (see Chapter 6: Case Study: Sucrose), and following with trehalose (see Chapter 7: Case Study: Trehalose), raffinose (see Chapter 8: Case Study: Raffinose), fructose polymers or fructans (see Chapter 9: Case Study: Fructans), polysaccharides in general (see Chapter 10: Case Study: Polysaccharides), starch (see Chapter 11: Case Study: Starch), and glycogen (see Chapter 12: Case Study: Glycogen), reserve polysaccharides of plants and cyanobacteria, respectively, and structural polysaccharides as cellulose (see Chapter 13: Case Study: Cellulose). This third part ends with the description of the main sugar-phosphates (see Chapter 14: Case Study: Sugar Phosphates) and nucleoside diphosphate-sugars (see Chapter 15: Case Study: Nucleotide Sugars) involved in carbohydrate metabolism in photosynthetic organisms.

To close, my deepest gratitude to my wife, Graciela Salerno, who has been of vital support throughout the months preceding the production of the book, and also of fundamental help with the editing and proofreading of the manuscript. Without her help this book would have not been possible. And to my daughter, Sheila Pontis, who has helped me throughout the writing of this manuscript with her knowledge of English, design, and communication. In addition, I would like also to express my sincere thanks to Alejandro Parise for his advice and council on the writing of Chapter 5, General Introduction to Mass Spectrometry and Nuclear Magnetic Resonance, and my special gratitude to Valentina Mariscotti for helping me with the translation, and to Gonzalo Caló and Cintia Pereyra who assisted me toward its completion.

Horacio G. Pontis, PhD
Professor Emeritus of Biochemistry,
Universidad Nacional de Mar del Plata, and
Emeritus Senior Investigator and Vice-President,
Applied Biological Research Foundation — FIBA,
Buenos Aires, Argentina

PART I

General Analytical Methods

CHAPTER 1

Determination of Carbohydrates Metabolism Molecules

Chapter Outline

1.1 Introduction

The present chapter describes a selection of the most commonly used methods for the analysis of different monosaccharides, such as glucose, fructose, galactose, mannose, amino sugars pentoses, and uronic acids, as well as for the determination of disaccharides (sucrose and trehalose).

Methods for Analysis of Carbohydrate Metabolism in Photosynthetic Organisms.
DOI: http://dx.doi.org/10.1016/B978-0-12-803396-8.00001-6

Finally, methods for the quantification of other compounds involved in the metabolism of carbohydrates, such as inorganic phosphate, ADP, UDP, and UDP-glucose, are included. The application of these methods is of value in any biochemical study dealing with carbohydrate metabolism of photosynthetic organisms. Particularly, some applications to enzyme studies are covered in Part III: Case Studies.

1.2 Determination of Reducing Sugars by the Somogyi–Nelson Method

Principle

This classical method is one of the most widely used for the quantitative determination of reducing sugars. Sugars (aldoses or ketoses) containing a free carbonyl group in a slightly alkaline medium are in equilibrium with their corresponding enediol, which has a high reduction potential. Under these conditions, when heated with alkaline copper tartrate (Somogyi's reagent), these sugars reduce the copper from the cupric to cuprous state. Since the cuprous ion can be reoxidized to cupric, arsenomolybdic acid (Nelson's reagent) is added to stabilize the cuprous oxide produced, which in turn, reduces the molybdic acid to molybdenum blue (Nelson, 1944; Somogyi, 1952).

Reagents

Prepare the stock solutions of the two components of the Somogyi's reagent (solutions A and B) and of Nelson's reagent (chromogenic reagent). Somogyi's reagent (obtained by mixing solutions A and B in the proportion indicated below) as well as the dilution of Nelson's reagent should be freshly prepared at the moment of use.

Somogyi's reagent stocks

Solution A

Na_2CO_3 (sodium carbonate anhydrous)	25 g
$KNaC_4H_4O_6 \cdot 4H_2O$ (potassium sodium tartrate tetrahydrate)	25 g
$NaHCO_3$ (sodium bicarbonate)	20 g
Na_2SO_4 (sodium sulfate anhydrous)	200 g

Preparation of solution A: Add salts, one by one, to a volume of 800 mL distilled water, under continuous stirring. Allow each salt to be completely dissolved before adding the next one. Make up to 1000 mL.

Storage conditions: Dark glass bottle, at 20–30°C.

Solution B

$CuSO_4 \cdot 5H_2O$ (copper sulfate pentahydrate)	15 g
H_2SO_4 (c) (sulfuric acid, δ: 1836 $kg \cdot L^{-1}$)	2 drops

Preparation of solution B: Dissolve the salt in 50 mL of distilled water, add the drops of acid while stirring and make up to 100 mL.

Storage conditions: Room temperature. Make up freshly every 2 months.

Nelson's reagent stock

$(NH_4)_6Mo_7O_2 \cdot 4H_2O$ (ammonium molybdate tetrahydrate)	25 g
H_2SO_4 (c) (sulfuric acid, δ: 1836 $kg \cdot L^{-1}$)	21 mL
$Na_2HAsO_4 \cdot 7H_2O$ (sodium arsenate dibasic heptahydrate)	3 g

Preparation of stock solution: Dissolve the ammonium molybdate in 450 mL of distilled water. Add the sulfuric acid while stirring. Separately, dissolve the arsenate in 25 mL of distilled water and add to the molybdate solution while mixing. Incubate the solution 24–48 h at 35–40°C.

Storage conditions: Dark glass bottle, at 35–40°C. Stable for at least one year. The reagent should be yellow without a green tinge.

Preparation of diluted solutions: Just prior to use, proceed to dilute the stock solutions as follows:

> *Somogyi's reagent dilution*: Mix solution A, solution B, and distilled water (25:1:26 v/v/v).
> *Nelson's reagent dilution*: Mix Nelson's solution stock and distilled water (1:2 v/v).

Note: Diluted working solutions are stable for 1 day.

Procedure

The sample to be analyzed and the standard solution containing between 10 and 100 nmol of the sugar should be in a 500 μL volume. Add 500 μL of the diluted Somogyi's reagent, mix, cap the tubes, and heat at 100°C for 20 min. After cooling in a water bath to room temperature, add 800 μL of the diluted Nelson's reagent and mix vigorously. Let stand for 1 min and measure absorbance at 520 or 660 nm (Fig. 1.1) against a reaction blank prepared by replacing the volume of sample with water. The deep blue color is stable for several hours.

Comments

To enhance the method sensitivity, the final sample volume can be reduced up to 50 μL; however, the volumes of the Somogyi's and Nelson's reagents should be maintained. Notice that sensitivity is four times higher when determining absorbance at 660 nm (Fig. 1.1).

Not all monosaccharides react with the same intensity to the Somogyi–Nelson reagent. Glucose and fructose develop equal color intensity, while mannose and galactose result in a less intense blue. Likewise, the aldopentoses (ribose, xylose) and ketopentoses (ribulose, xylulose) show lower color intensity than their respective hexoses.

1.3 Determination of Total Sugars by the Anthrone Method

Principle

This method allows quantification of hexoses, aldopentoses, uronic acids, and neutral hexoses present in a glycoprotein, without previous hydrolysis. The simplicity of the procedure coupled

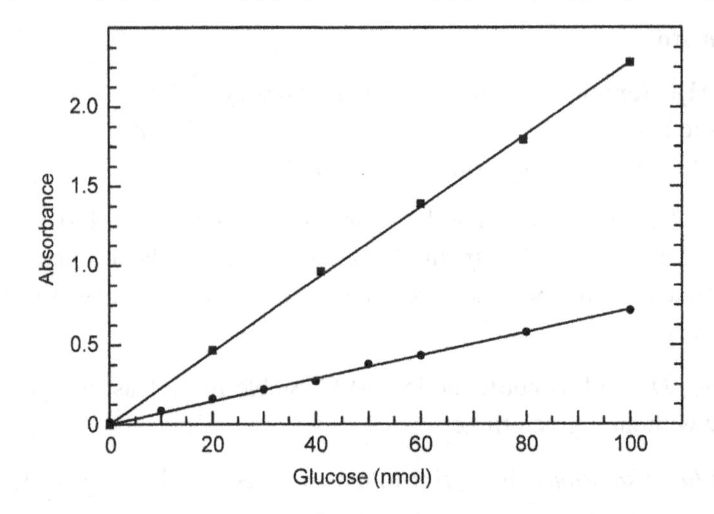

Figure 1.1
Standard curve for glucose determination by the Somogyi–Nelson method. Absorbance was measured at 520 nm (●) and 660 nm (■).

with its great sensitivity is one of the outstanding features that this assay offers, being one of the most successfully applied for carbohydrate determination (Ashwell, 1957).

In the presence of sulfuric acid–anthrone reagent, carbohydrates are dehydrated to furfural (or hydroxymethylfurfural), which in turn condenses with anthrone (9,10-dihydro-9-oxoanthracene) to produce a bluish-green complex (Dreywood, 1946). The intensity of the color is quantified by measuring absorbance at 620 nm.

Reagents

H_2SO_4 (c) (sulfuric acid, δ: 1836 kg · L^{-1})	72 mL
Anthrone	50 mg
Thiourea	1 g

Preparation of anthrone reagent: Add 72 mL of sulfuric acid to 28 mL of distilled water. While the mix is still warm, add the anthrone and thiourea while stirring until dissolution. Let stand for 4 h. The solution features an intense yellow color.

Storage conditions: Dark glass bottle, refrigerated at 4°C. The reagent is stable for at least two weeks under refrigeration. The instability of the reagent has been pointed out as a disadvantage.

Procedure

Add 0.8 mL of cold anthrone reagent to 100 μL of sample (or a standard solution containing between 20 and 200 nmol of hexoses). After vigorous mixing, cap the tubes and heat at 100°C for 15 min. Cool the tubes in a water bath to room temperature for 20 min and

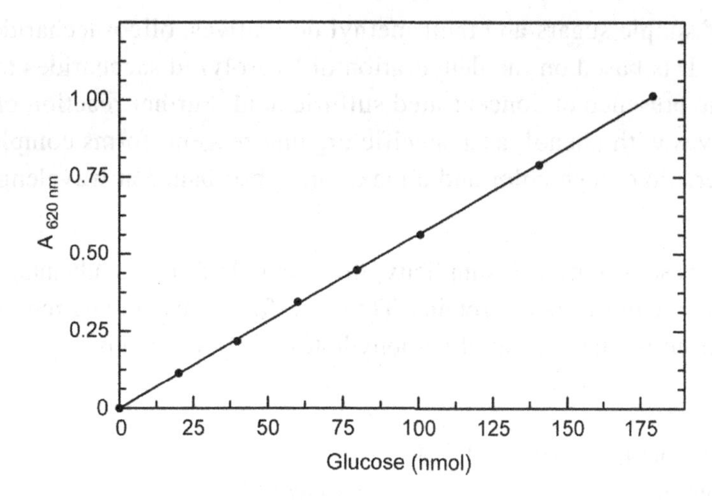

Figure 1.2
Standard curve for glucose determination by the anthrone method.

determine absorbance at 620 nm (Fig. 1.2) against a reaction blank prepared by replacing the volume of sample with water.

Comments

The size of the glass tube used in the assay must allow a vigorous mixing of the reagents and ensure that no loss of liquid reaction occurs.

Galactose, mannose, glucose, and fucose, present in glycoproteins, strongly react with this reagent. Despite the fact that the color developed by the sugars in the reaction is the same, the intensity varies per mole of monosaccharide. The color developed by the different sugars decreases as follows: glucose = fructose > galactose > mannose > fucose. The color yield for fucose is about threefold lower than that of glucose. Pentoses are considerably less sensitive and exhibit a maximum absorbance at a wavelength of 600 nm. The reaction is also positive with all mono-, di-, and polysaccharides, as well as with dextrins, starches, and plant polysaccharides.

Samples should be free of tryptophan-rich proteins to avoid interference in color development.

1.4 Determination of Total Sugars by the Phenol–Sulfuric Acid Assay

Principle

Among the colorimetric methods available for carbohydrate analysis, this is the most reliable one (Dubois et al., 1956). This rapid and simple procedure allows the determination of simple sugars and their methyl derivatives, oligosaccharides, and polysaccharides. It is based on the dehydration of hydrolyzed saccharides to furfural derivatives in the presence of concentrated sulfuric acid. Further reaction of the furfural derivatives with phenol, as a specific organic reagent, forms complexes with its characteristic orange color and a maximum absorbance at wavelength of 490 nm.

Notwithstanding its sensitivity and simplicity, this method offers the advantage of being hardly affected by the presence of proteins. Thus, it affords a useful and reproducible technique for measuring the content of carbohydrates in glycoproteins.

Reagents

H_2SO_4 (c) (sulfuric acid, δ: 1836 kg · L^{-1})
Phenol (reagent grade) 80% (w/w)

Preparation of phenol reagent: Add 20 mL of distilled water to 80 g of phenol.

Storage conditions: The phenol reagent is stable for several months at room temperature. A pale yellow color usually develops upon solution storage. This color does not interfere with the assay.

Procedure

To 500 μL of sample (or a standard solution containing between 10 and 100 μg of glucose) add 500 μL of phenol reagent, followed by a rapid addition of 2.5 mL of concentrated sulfuric acid. Let stand at room temperature for 15 min. Allow the solution to cool and measure absorbance at 480–490 nm. Read samples against a reaction blank containing distilled water instead of sugar solution. The orange color is stable for several hours.

Comments

To facilitate heat dissipation resulting from the addition of sulfuric acid to the aqueous solution, it is convenient to use glass wide-mouth tubes (eg, 16–20 mm in diameter).

Hexoses, di-, oligo-, and polysaccharides, including their methylated derivatives, with free or potentially free reducing groups, react with the phenol reagent and develop a yellow-orange color reaching the maximum absorption at 485–490 nm. Pentoses, methyl pentose, and uronic acids also react and develop a color that has maximum absorbance at 480 nm. For certain compounds such as xylose, the sensitivity of the reaction can be increased by halving the amount of phenol in the reaction mixture.

A recent method based on UV spectrophotometry avoids the use of phenol, and the absorption inherent to furfural derivatives is directly measured at a wavelength of 315 nm (Albalasmeh et al., 2013).

1.5 Enzymatic Determination of Glucose and Fructose

Principle

This is a rapid and specific method based on the enzymatic conversion of hexoses to phosphogluconic acid in the presence of $NADP^+$ (oxidized form of nicotinamide adenine dinucleotide phosphate) which is reduced to NADPH (molar extinction coefficient at 340 nm = 6.220 $M^{-1}cm^{-1}$) (Sturgeon, 1980). The difference in the ultraviolet absorption spectrums between the $NADP^+$ and NADPH makes it simple to measure the conversion of one to another by measuring absorbance at 340 nm using a spectrophotometer.

In a first step of the enzymatic coupled reactions, the hexoses (fructose and glucose) are phosphorylated by hexokinase. Then the fructose-6-phosphate produced is isomerized to glucose-6-phosphate in the presence of phosphoglucose isomerase. Finally, the resulting glucose-6-phosphate is oxidized to phosphogluconate in the presence of $NADP^+$ and glucose-6-phosphate dehydrogenase. The amount of NADPH produced is quantified by measuring the changes in absorbance at 340 nm and is proportional to the amount of hexoses present in the sample.

The enzymatic reactions can be summarized as follows:

$$\text{Fructose} + \text{ATP} \xrightarrow{\text{Hexokinase}} \text{Fructose-6-phosphate} + \text{ADP}$$

$$\text{Glucose} + \text{ATP} \xrightarrow{\text{Hexokinase}} \text{Glucose-6-phosphate} + \text{ADP}$$

$$\text{Fructose-6-phosphate} \xleftrightarrow{\text{Phosphoglucose isomerase}} \text{Glucose-6-phosphate}$$

$$\text{Glucose-6-phosphate} + \text{NADP}^+ \xrightarrow{\text{Glucose-6-phosphate dehydrogenase}} \text{6-Phosphogluconate} + \text{NADPH} + \text{H}^+$$

Reagents

ATP	100 mM
$MgCl_2$ (magnesium chloride)	200 mM
$NADP^+$	100 mM
Tris-HCl buffer (pH 8.0)	1 M
Hexokinase (from yeast)	250 U · mg protein^{-1}
Phosphoglucose isomerase (from yeast)	400 U · mg protein^{-1}
Glucose-6-phosphate dehydrogenase (from yeast)	200 U · mg protein^{-1}

Procedure

To 50 μL of sample (or a standard solution containing between 5 and 100 nmol of glucose and/or fructose), add 6 μL of 100 mM ATP, 6 μL of 200 mM $MgCl_2$, 6 μL of 100 mM

$NADP^+$, 10 μL of 1 M Tris-HCL buffer (pH 8), 5 μL of hexokinase (2 μg), 5 μL of phosphoglucose isomerase (1 μg), 5 μL of glucose-6-phosphate dehydrogenase (1 μg), and 7 μL of water (100 μL total reaction volume).

Incubate at 37°C for 10 min. Make up to 1 mL with distilled water, mix and determine absorbance at 340 nm. To calculate the μmoles of hexoses present in the sample, divide the increase in absorbance by 6.22, taking into account that the absorbance of 1 mL containing 1 μmol of NADPH, in a 1-cm path length cuvette at 340 nm is 6.22.

To determine only glucose, add the same amount of reagents to a 50-μL sample but omit the addition of phosphoglucose isomerase, and replace its volume with 5 μL of distilled water. The procedure is completed as described above for a mixture of glucose and fructose.

Comments

Before proceeding with an unknown sample, it is recommended to check with a hexose standard solution that the enzymatic reactions are fully carried out and that all the sugar is being quantified.

This method can also be used for mannose, fructose, and glucose determination in the same assay mixture. Based on the same principle, mannose is phosphorylated by hexokinase, and the mannose-6-phosphate produced is converted to fructose-6-phosphate by adding phosphomannose isomerase (Sturgeon, 1980). Similarly, sugar phosphates (fructose-6-phosphate, and/or glucose-6-phosphate, and/or mannose-6-phosphate) can be determined omitting hexokinase in the incubation mixture, and incubating with the corresponding auxiliary enzymes.

1.6 Enzymatic Determination of Sucrose

After sucrose hydrolysis by invertase, the glucose and fructose produced can be measured by the enzymatic assay detailed in Section 1.5. Jones et al. (1977) proposed alternative and extremely sensitive methods to determine NADPH. Thus, sucrose and its hydrolysis products can be quantified at a concentration in the order of picomol. The principle used to achieve this sensitivity can be extended to determine other disaccharides such as trehalose. In this case, the method allows quantifying the glucose molecules released after hydrolysis with trehalase.

Principle

Sucrose is an alkali stable saccharide that can be specifically quantified in the presence of free monosaccharides (particularly, glucose and fructose), which can be destroyed by heating in the presence of sodium hydroxide. Sucrose is then determined after the action of four auxiliary enzymes in a single analytical step, as follows:

$$\text{Sucrose} \xleftrightarrow{\text{Invertase}} \text{Glucose} + \text{Fructose}$$

$$\text{Glucose} + \text{Fructose} + 2\,\text{ATP} \xrightarrow{\text{Hexokinase}} \text{Glucose-6-phosphate} + \text{Fructose-6-phosphate} + 2\,\text{ADP}$$

$$\text{Fructose-6-phosphate} \xleftrightarrow{\text{Phosphoglucose isomerase}} \text{Glucose-6-phosphate}$$

$$2\,\text{Glucose-6-phosphate} + 2\,\text{NADP}^{+} \xrightarrow{\text{Glucose-6-phosphate dehydrogenase}} \text{6-Phosphogluconate} + 2\,\text{NADPH}$$

The NADPH produced in the last reaction can be quantified by different procedures, depending on the amount produced. The most sensitive methods were addressed in detail by Jones et al. (1977). Three of them are specified below.

1.6.1 Spectrophotometric Method (Measurement of NADPH Absorption)

Reagents

ATP	100 mM
$MgCl_2$	200 mM
$NADP^+$	100 mM
Imidazole–HCl buffer (pH 6.9)	0.5 M
Dithiothreitol (DTT)	50 mM
Bovine serum albumin (BSA)	2% (w/v)
Invertase (from yeast)	$300\ U \cdot mg\ protein^{-1}$
Hexokinase (from yeast)	$250\ U \cdot mg\ protein^{-1}$
Phosphoglucose isomerase (from yeast)	$400\ U \cdot mg\ protein^{-1}$
Glucose-6-phosphate dehydrogenase (from yeast)	$200\ U \cdot mg\ protein^{-1}$

Procedure

Bring the sugar sample to 0.02 N NaOH and heat at 95°C for 30 min. To 200 μL of the heated sample (or a standard solution containing between 5 and 70 nmol of sucrose), add 0.8 mL of a mix consisting of 100 mM imidazole–HCl buffer (pH 6.9), 0.4 mM $NADP^+$, 1 mM ATP, 5 mM $MgCl_2$, 0.5 mM DTT, 0.02% (w/v) BSA, $20\ \mu g \cdot mL^{-1}$ yeast invertase, $2\ \mu g \cdot mL^{-1}$ yeast hexokinase, $1\ \mu g \cdot mL^{-1}$ yeast phosphoglucose isomerase, and $1\ \mu g \cdot mL^{-1}$ yeast glucose-6-phosphate dehydrogenase. After incubation at 20–25°C for 30 min, measure absorbance at 340 nm.

1.6.2 Direct Fluorometric Method (Measurement by NADPH Native Fluorescence)

Procedure

This assay is performed as outlined in Section 1.6.1 (spectrophotometric method) except that the enzymatic reaction is carried out in a fluorometer using a 3-mL (10 × 75 mm) tube and the reagent mix is modified as follows: the imidazole buffer is reduced to 50 mM, ATP to 0.2 mM, and $NADP^+$ to 0.1 mM. For the 0.1–0.5 nmol range, the $NADP^+$ used is 0.03 mM to keep the fluorescence blank to a minimum. The measurement by NADPH native fluorescence allows a range of detection between 0.1 and 5 nmol of sucrose.

1.6.3 Assay With Fluorescence Enhancement

Reagents

Phosphate buffer (0.25 M Na_3PO_4, 0.25 M K_2HPO_4)	0.25 M
NaOH	10 N
H_2O_2 (hydrogen peroxide)	40 mM

Preparation of NaOH–H_2O_2 reagent: Just prior to use, proceed to prepare a 6 N NaOH solution containing 10 mM hydrogen peroxide.

Procedure

The first step is identical to that in the previous method (Section 1.6.2). Yet to achieve greater sensitivity, the reaction volume is reduced. An aliquot of 10 μL (or a smaller volume) could be heated in small sealed tubes (or under oil in oil wells) to prevent evaporation. The reagent mix is the same as in Section 1.6.2. Add a 10-μL (or less) sample to 50 μL of the mix in a 3-mL fluorometer tube. After incubation at 20–25°C for 30 min, add 50 μL of phosphate buffer (0.25 M Na_3PO_4, 0.25 M K_2HPO_4) and mix thorough and carefully to avoid splashing. After heating for 15 min at 60°C, add 1 mL of 6 N NaOH containing 10 mM H_2O_2 and 10 mM imidazole, and immediately mix (Lowry and Carter, 1974). Heat the tubes once again at 60°C for 15 min. Finally, cool the tubes to room temperature and read fluorescence. Fluorescence is virtually indefinitely stable. The measurement by NADPH by the greater fluorescence when converted back to $NADP^+$ with H_2O_2 and treated with NaOH allows a range of detection between 0.01 and 1 nmol of sucrose.

Comments

The initial heating in NaOH allows destroying not only free hexoses, but also the enzymes present in the tissue which might interfere.

Since the enzymatic reactions have different pH optima, and invertase pH optimum is approximately 4.5, the pH 6.9 for the reaction mixture is a compromise. It is recommended to check each invertase preparation and adjust the amount before use by performing a time course of the reaction with a standard sucrose solution.

1.7 Determination of Glucose and Galactose by an Enzymatic Colorimetric Method

Principle

This method is based on the specific oxidation of glucose (or galactose) by the enzyme glucose oxidase (or galactose oxidase), to give the corresponding lactone and hydrogen peroxide. Dahlqvist (1961) devised a coupled enzymatic method using peroxidase to catalyze the oxidation of the peroxide with the concomitant production of a colored compound which can be colorimetrically determined. Since the donor molecule originally used (2-dianisidine) has a high toxicity, it was replaced by a number of less hazardous substances. A variant is that the hydrogen peroxide, in the presence of peroxidase, oxidizes the chromogenic compound 2,2′-azino-di-(3-ethyl benzythiazolin-6-sulfonic acid) (ABTS, a stable, nontoxic, and water-soluble dye substrate). The absorbance of the brilliant blue-green colored product (oxidized ABTS) is measured at 440 nm (Okuda et al., 1977).

The enzymatic reactions are as follows:

$$\text{Glucose} + O_2 + H_2O \xrightarrow{\text{Glucose oxidase}} \text{Gluconic acid} + H_2O_2$$

$$H_2O_2 + \text{Donor}_{\text{reduced}}(\text{ABTS}_{\text{red}}) \xrightarrow{\text{Peroxidase}} \text{Donor}_{\text{oxidized}}(\text{ABTS}_{\text{ox}}) + H_2O$$

Reagents

Tris-HCl buffer (pH 7.0)	0.1 M
Glucose oxidase (or 20 mg, 20 $U \cdot mg^{-1}$ galactose oxidase)	10 mg, 40 $U \cdot mg^{-1}$
Horseradish peroxidase (HRP)	5 mg, 250 $U \cdot mg^{-1}$
ABTS (2,2′-azino-di-(3-ethyl benzythiazolin-6-sulfonic acid)	5 mg
HCl (hydrochloric acid)	5 N

Preparation of reagent: Add glucose oxidase (or galactose oxidase), peroxidase and ABTS to 50 mL of Tris-HCl buffer (pH 7.0). Make up to 100 mL with distilled water. The resulting solution is stable at 4°C for several weeks.

Procedure

Add 100 μL of sample (or a standard solution containing between 5 and 250 nmol of glucose (or galactose)) to 2.9 mL of the reagent containing ABTS. Incubate at 37°C for 30 min for glucose determination (or 60 min at the same temperature for galactose determination). Stop the reaction by adding 100 μL of 5 N HCl. Let stand for 5 min and measure absorbance at 440 nm.

1.8 Determination of Glucose by an Enzymatic Fluorometric Method

Principle

This method is based on the same principle as that described in Section 1.7, ie, the specific oxidation of glucose by the enzyme glucose oxidase producing D-glucono-1,5-lactone and hydrogen peroxide (Dahlqvist, 1961). The fluorogenic compound N-acetyl-3,7-dihydroxyphenoxazine (ADHP, commercially known as Amplite and Amplex Red) is the most sensitive and stable probe used to detect hydrogen peroxide in the presence of peroxidase, producing the highly fluorescent compound resorufin. The reaction stoichiometry of ADHP and H_2O_2 was determined to be 1:1 (Zhou et al., 1997). Resorufin has an excitation maximum at ~563 nm, an emission maximum at ~587 nm, and a molar extinction coefficient of 54,000 $cm^{-1}M^{-1}$. Consequently, the assay can be carried out either fluorometrically or spectrophotometrically.

Reagents

Phosphate buffer (pH 7.4)	10 mM
HRP	$10\ U \cdot mL^{-1}$
Glucose oxidase	$100\ U \cdot mL^{-1}$
10-acetyl-3,7-dihydroxyphenoxazine (ADHP, diluted in DMSO)	10 mM
Dimethylsulfoxide (DMSO)	

Preparation of ADHP stock solution: Prepare the stock solution (10 mM) in analytically pure DMSO. Divide the slightly red solution into small aliquots and store at −20°C in the dark, for up to 6 months. Immediately before use, thaw an aliquot of stock solution. Since ADHP is an air sensitive reagent, use it promptly and protected from light. ADHP is unstable in the presence of thiols (such as dithiothreitol) and 2-mercaptoethanol at concentrations higher than 10 μM, and at a pH>8.5.

Preparation of the reaction mixture: The reaction mixture consists of 100 μM ADHP, $0.2\ U \cdot mL^{-1}$ peroxidase, and $2\ U \cdot mL^{-1}$ glucose oxidase. To prepare 5 mL of this solution, mix 50 μL of the stock solution of 10 mM ADHP, 100 μL of HRP $10\ U \cdot ml^{-1}$, 100 μL of glucose oxidase $100\ U \cdot mL^{-1}$, and 4.75 mL of 10 mM phosphate buffer (pH 7.4).

Procedure

The assay is designed to be conducted in microplates. Add 50 μL of the reaction mixture containing ADHP, HRP, and glucose oxidase to each well containing 50 μL of sample (or a standard solution containing between 10 and 100 nmol of glucose), and no-glucose control.

Incubate the reaction at room temperature for 30 min protected from light. Determine fluorescence or absorbance in a reader equipped for excitation in the range of 530–560 nm and fluorescence emission detection at ~587 nm, or for absorbance at ~560 nm. The assay is continuous and, therefore, fluorescence or absorbance may be measured at different points in time to follow the kinetics of the reactions.

Comments

This procedure can be adapted for use with a standard fluorometer, increasing volumes on the basis of the cuvette capacity.

1.9 Determination of Amino Sugars by the Elson–Morgan Method

Principle

Amino sugars (eg, glucosamine (2-amino-2-deoxy-D-glucose) and galactosamine (2-amino-2-deoxy-D-galactose)) usually occur in nature as N-acetyl derivatives integrating the glycan chain of glycoproteins. After acid hydrolysis, these sugars are released as N-deacetylated monosaccharides (2-amino-2-deoxy-hexoses), which can be determined by a colorimetric procedure (Elson and Morgan, 1933). This method is based on the condensation of a free 1-aldo-2-amino sugar with acetylacetone by heating in alkaline solution. The product is then allowed to react with 4-(N,N-dimethylamino) benzaldehyde in acid, to yield a red product which can be quantified by measuring absorbance at 530 nm.

Reagents

Reagent 1

Acetylacetone (pentane-2-4-dione) redistilled	2 mL
Sodium carbonate	1 M

Preparation of Reagent 1: Dissolve 2 mL of acetylacetone in 100 mL of 1 M sodium carbonate. Prepare freshly just before use.

Reagent 2

4-(N,N-dimethylamino) benzaldehyde	0.8 g
Ethanol	
HCl (c) (fuming hydrochloric acid, 37%, ACS reagent)	

Preparation of Reagent 2: Dissolve 0.8 g of 4-(N,N-dimethylamino) benzaldehyde in 30 mL of ethanol and 30 mL of HCl.

Procedure

Add 250 μL of Reagent 1 to a 250-μL sample (or a standard solution containing up to 500 nmol of 2-amino hexoses) in a stoppered tube and heat at 100°C for 20 min. After cooling to room temperature, add 1.75 mL of ethanol and 1 mL of Reagent 2, followed by thorough mixing. Incubate at 65°C for 10 min. Cool the tubes to room temperature and measure absorbance at 530 nm.

Comments

This procedure is recommended to determine isolated amino sugars. In the case of crude hydrolyzates, neutral sugars and amino acids could interfere, giving orange colored products.

Samples from hydrochloric acid hydrolysis of glycoproteins must be dried by evaporation in vacuum. Final traces of acid can be removed by storing the samples in a vacuum desiccator in the presence of NaOH pellets for at least 18 h.

1.10 Determination of Acetyl Amino Sugars by the Morgan–Elson Method

Principle

After acid hydrolysis, acetyl amino sugars are released from glycoproteins as 2-acetamido-2-deoxy-hexoses which can be determined by the Morgan–Elson method (Morgan and Elson, 1934) modified by Reissig et al. (1955) in order to increase sensitivity.

Reagents

Reagent 1	
Potassium tetraborate	0.2 M
Reagent 2	
4-(N,N-dimethylamino) benzaldehyde	
Glacial acetic acid	
HCl	10 N

Preparation of Reagent 2 stock: Dissolve 10 g of 4-(N,N-dimethylamino) benzaldehyde in 100 mL glacial acetic acid-10 M HCl (7:1, v/v).

Storage conditions: Dark glass bottle. Stock solution is stable for several weeks.

Working solution: Immediately before use, dilute the reagent (1:9, v/v) with glacial acetic acid.

Procedure

To a 250-μL sample (or a standard solution containing up to 25 nmol of 2-acetamido-2-deoxyhexose) add 50 μL of Reagent 1 and heat at 100°C for 3 min. After cooling to room temperature, add 1.5 mL of Reagent 2 followed by thorough mixing and incubation at 37°C for 20 min. Cool to room temperature and determine absorbance at 585 nm.

1.11 Determination of Uronic Acids by the Carbazole Assay

Principle

The analysis of uronic acids can be conducted by conversion to chromogenic compounds (furfural type) after being heated in the presence of strong acid (Dische, 1949).

Reagents

Reagent 1

Sodium tetraborate	0.025 M in sulfuric acid
H_2SO_4 (c) (sulfuric acid, δ: 1836 $kg \cdot L^{-1}$)	

Preparation of Reagent 1 stock: Add 98 mL of concentrated sulfuric acid to sodium tetraborate (0.95 g in 2.0 mL of distilled water). The resulting solution is stable at 4°C.

Reagent 2

Carbazole (ethanol recrystallized) 0.125% (w/v, in ethanol)

Preparation of Reagent 2 stock: Dissolve 125 g of carbazole in 100 mL of ethanol.

Procedure

Pre-cool samples and reagents to 4°C. Add 1.5 mL of Reagent 1 to a 250-μL sample (or a standard solution containing up to 200 nmol of uronic acid) with mixing and cooling. Heat the mixture for 10 min at 100°C. After rapidly cooling at 0–4°C, add 50 μL of Reagent 2 and mix. Heat at 100°C for 15 min. After cooling, determine absorbance at 525 nm. The pink color produced is stable for about 16 h.

Comments

Hexoses and pentoses interfere in the assay, although pentoses to a lesser extent.

1.12 Determination of Pentoses by the Orcinol Procedure

Principle

The pentoses heated in a strong acid medium give furfural, which reacts with orcinol generating a colored compound (Brown, 1946).

Reagents

Reagent 1	
Trichloroacetic acid (TCA)	10% (w/v)
Reagent 2	
Orcinol (reagent grade, recrystallized if necessary)	0.2% (w/v, in 9.6 N HCl)
HCl (c) (fuming hydrochloric acid, 37%, ACS reagent)	
$NH_4Fe(SO_4)_2$ (ferric ammonium sulfate)	1.15% (w/v, in 9.6 N HCl)

Preparation of Reagent 2 stock: Prepare 0.2% (w/v) orcinol solution and 1.15% (w/v) ferric ammonium sulfate in 9.6 N HCl. Do it freshly just before use.

Procedure

Mix a 200-μL sample (or a standard solution containing 25–250 nmol of pentoses) with 200 μL of Reagent 1 and heat the mixture at 100°C for 15 min. After cooling to room temperature, add 1.2 mL of Reagent 2. After mixing, heat at 100°C for 20 min and then cool to room temperature. Determine absorbance at 660 nm. The developed color is stable for several hours.

Comments

Hexoses do not interfere with the determination of pentoses.

1.13 Determination of Fructose Derivatives by the Thiobarbituric Acid Assay

Principle

When ketoses are heated in an acid medium they become furfural derivatives which react with the thiobarbituric acid producing a yellow compound (Percheron, 1962). This method allows to determine sucrose (alkali stable at 100°C) in the presence of free monosaccharides by quantifying its fructose moiety.

Reagents

Thiobarbiturc acid (TBA)	342 mg
HCl (c) (fuming hydrochloric acid, 37%, ACS reagent)	
HCl	0.01 N
NaOH	0.5 N

Preparation of TBA reagent stock: Dissolve 342 mg of TBA in 100 mL of 0.01 N HCl.

Storage conditions: The TBA stock solution is stable when it is stored in a dark bottle at 0–4°C for at least a month.

Working solution (TBA-HCl): Half dilute TBA stock reagent with HCl (c) (1:1, v/v), to obtain the TBA-HCl working reagent. Prepare freshly just before use.

Procedure

To determine free fructose, mix 50 μL of sample (or a standard solution containing between 5 and 80 nmol of fructose) with 200 μL of water and add 600 μL of TBA-HCl. Heat at 100°C for 7 min. After cooling to room temperature in a water bath, measure absorbance at 432 nm.

To determine sucrose, add 200 μL of 0.5 N NaOH to a 50-μL sample (or a standard solution containing between 5 and 80 nmol of sucrose) and heat at 100°C for 10 min. After cooling

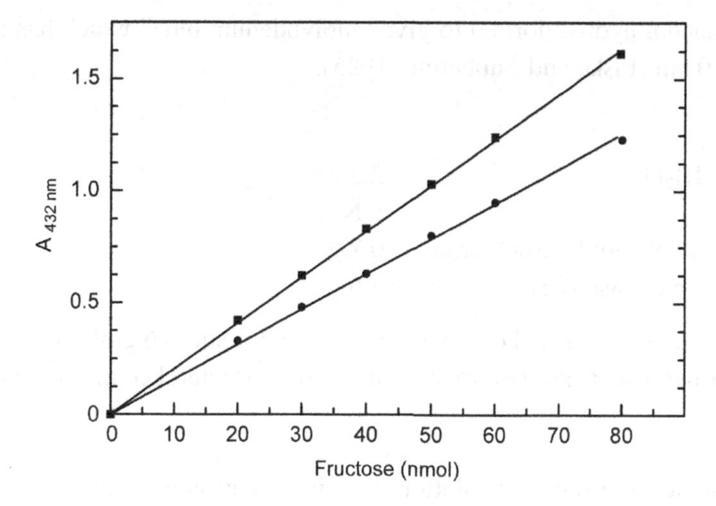

Figure 1.3
Standard curve for the determination of fructose (■) and fructose-6-phosphate (●) by the thiobarbituric acid assay.

to room temperature, add 600 μL of TBA-HCl, and heat at 100°C for 7 min. After cooling, measure absorbance at 432 nm (Fig. 1.3).

Comments

This method is applicable to all sugars containing fructose, such as sucrose, sucrose-6-phosphate, raffinose, and fructose polymers (fructans). Phosphorylated sugars containing fructose develop a less intense color per unit mass as compared to the respective nonphosphorylated sugars.

In general, commercial TBA is not pure enough to carry out this reaction and the reagent blank could give a high intensity color. As a consequence, the chemical should be purified before preparing the reagent. It is recommended to perform the procedure under low light intensity, avoiding direct light exposure. TBA purification is as follows: 8 g of TBA are dissolved in 220 mL of distilled water previously heated to 80–85°C. The solution is filtered through a folded paper filter, and the filtrate is loaded onto an alumina column (neutral aluminum oxide) 3 × 4.1 cm diameter. The eluate is cooled to room temperature and then lyophilized. The resulting powder should be stored in a dark bottle at −20°C, and remains stable under these conditions for at least 1 year.

1.14 Determination of Inorganic Phosphate by the Fiske–Subbarow Assay

Principle

This is a colorimetric assay designed to measure levels of inorganic phosphate by the formation of phosphomolybdic acid, which is then reduced by a reducing agent (eg, 4 diamino phenol hydrochloride) to give "molybdenum blue" which has a maximum absorption at 660 nm (Fiske and Subbarow, 1925).

Reagents

$(NH_4)_6Mo_7O_{24} \cdot 4H_2O$	2.5%
Sulfuric acid	5 N
Amidol (4 diamino phenol hydrochloride)	0.2 g
$Na_2S_2O_5$ (sodium metabisulfite)	3.6 g

Preparation of reducer reagent: Dissolve 0.2 g of amidol and 3.6 g of sodium metabisulfite in 20 mL of distilled water. Prepare freshly just before use and keep in the dark.

Procedure

To a 500-μL sample (or a standard solution containing between 0.1 and 0.5 μmol of inorganic phosphate) add 100 μL of 5 N sulfuric acid, 100 μL of 2.5% ammonium molybdate, and 200 μL of the reducer reagent. After mixing, let stand for 10 min at room temperature and determine absorbance at 660 nm (Fig. 1.4).

Comments

Organic phosphates can also be determined by this method, after appropriate acid hydrolysis conditions that depend on the lability of the bound phosphate (Leloir and Cardini, 1957). Accordingly, organophosphates can be grouped into: (1) extra labile phosphates (eg, phosphocreatine, 1,3-diphosphoglyceric acid, ribose-1-phosphate, deoxyribose-1-phosphate, fructose-2-phosphate, and acyl phosphates); (2) labile phosphates

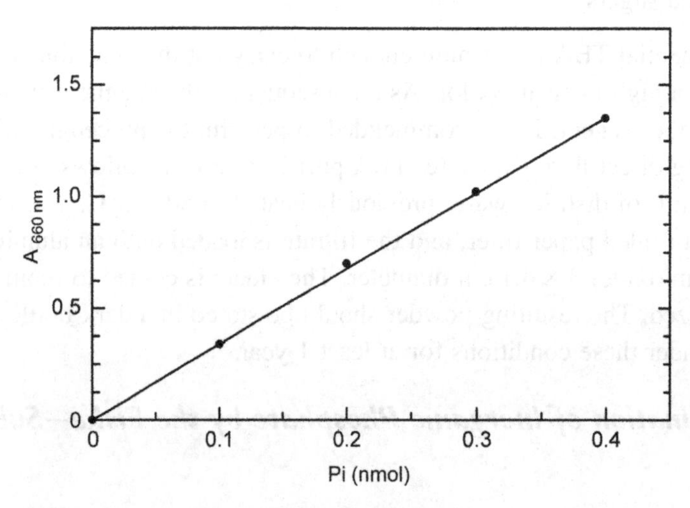

Figure 1.4
Standard curve for inorganic phosphate determination by the Fiske–Subbarow assay.

(eg, adenosine triphosphate, adenosine diphosphate, uridine diphosphate, aldose-1-phosphate, fructose-1-phosphate, fructose-1,6-diphosphate, and glucuronic acid 1-phosphate); and (3) stable phosphates (eg, phosphopyruvic acid, hexose-6-phosphate, pentose-3-phosphate, pentose-5-phosphate and the corresponding mono- and di-nucleotides, 6-phosphogluconic acid, glycerol and glycerophosphates, inositol phosphates, phosphorylcholine, and phosphorylethanolamine).

1.15 Determination of Inorganic Phosphate by the Chifflet's Method

Principle

This is an improved method of the original Fiske–Subbarow assay (Fiske and Subbarow, 1925) described in Section 1.14. The modifications introduce several advantages, such as higher sensitivity and the possibility of determining inorganic phosphate in the presence of labile organic phosphates and high protein concentration (Chifflet et al., 1988).

Reagents

Sodium dodecyl sulfate (SDS)	7.2%
Ascorbic acid	6% (in HCl 1 N)
HCl	1 N
$(NH_4)_6Mo_7O_{24} \cdot 4H_2O$	1%
Sodium citrate	2%
Acetic acid	2%
$NaAsO_2$ (sodium arsenite)	2%

Reagent A

7.2% SDS

Reagent B

Solution (1): 6% (w/v) ascorbic acid (prepared in 1 N HCl). Immediately after preparing, place on ice and in the dark. The solution remains useful for 10 min.

Solution (2): 1% ammonium molybdate.

Preparation of Reagent B: Mix equal volumes of solution (1) and solution (2). Prepare freshly just before use and keep in the dark.

Reagent C

Preparation of Reagent C: Mix equal volumes of 2% sodium citrate, 2% acetic acid, and 2% sodium arsenite. Prepare freshly just before use.

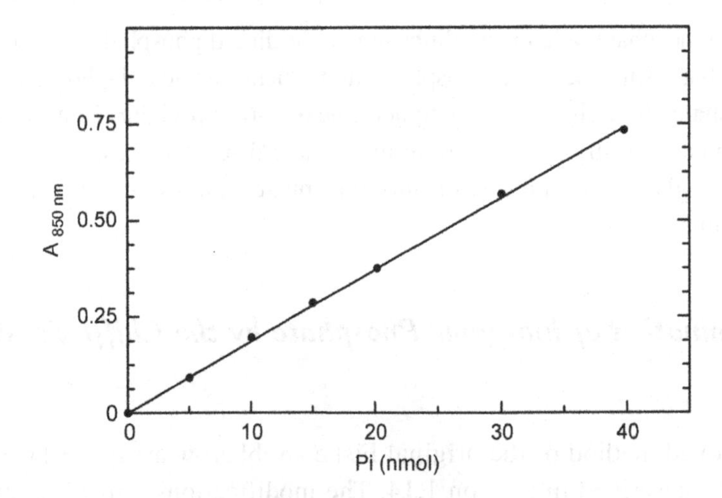

Figure 1.5
Standard curve for inorganic phosphate determination by the Chifflet's method.

Procedure

Add 250 μL of Reagent A and 300 μL of Reagent B to a 50-μL sample (or a standard solution containing between 1 and 50 nmol inorganic phosphate). Let the reaction mixture stand for 3 min and add 450 μL of reagent C. Let stand for 20 min and measure absorbance at 850 nm (Fig. 1.5).

Comments

This method is highly reproducible and the color remains stable for at least 3.5 h. Of special interest is its high tolerance to protein (up to 50 mg $\cdot$ mL^{-1} of protein in the sample without precipitation or color interference). Moreover, when it is used to measure an enzyme activity that releases inorganic phosphate, the addition of SDS allows stopping the reaction without organic phosphate hydrolysis.

1.16 Determination of Uridine Diphosphate by the Measurement of Pyruvate

Principle

The reaction catalyzed by the enzyme pyruvate kinase:

$$\text{UDP} + \text{Phosphoenolpyruvate} \leftrightarrow \text{UTP} + \text{Pyruvate}$$

can be used to analyze UDP by measuring the pyruvate formed either as 2.4-dinitrophenylhydrazone (Pontis and Leloir, 1962) or as the NADH oxidation after incubation with lactic dehydrogenase.

1.16.1 Colorimetric Method Using 2,4-Dinitrophenylhydrazine

Principle

The reaction between 2,4-dinitrophenylhydrazine and a ketone (such as pyruvate) produces a colored compound (2,4-dinitrophenylhydrazone), which is determined at 520 nm.

Reagents

KCl (potasium chloride)	0.4 M
Phosphoenolpyruvate (cyclohexylammonium or sodium salt)	0.01 M (prepared in 0.4 M KCl)
$MgSO_4$ (magnesium sulfate)	0.1 M
HCl	2 N
Dinitrophenylhydrazine	0.1% (prepared in 2 N HCl)
Tris-HCl buffer (pH 7.5)	1 M
Pyruvate kinase (from rabbit muscle)	350–500 U · mg protein^{-1}
Ethanol	95%
Sodium hydroxide	10 N

Procedure

To a 50-μL sample (or a standard solution containing between 10 and 150 nmol UDP) add 25 μL of phosphoenolpyruvate-KCl, 25 μL of pyruvate kinase (diluted in 0.1 M magnesium sulfate), and 10 μL of 1 M Tris-HCl buffer (pH 7.5). After incubating at 37°C for 15 min, stop the reaction by adding 150 μL of 0.1% dinitrophenylhidrazine (in 2 N HCl). After 5 min, add 200 μL of 10 N NaOH and 500 μL of ethanol. Mix and centrifuge. Determine absorbance at 520 nm (Fig. 1.6).

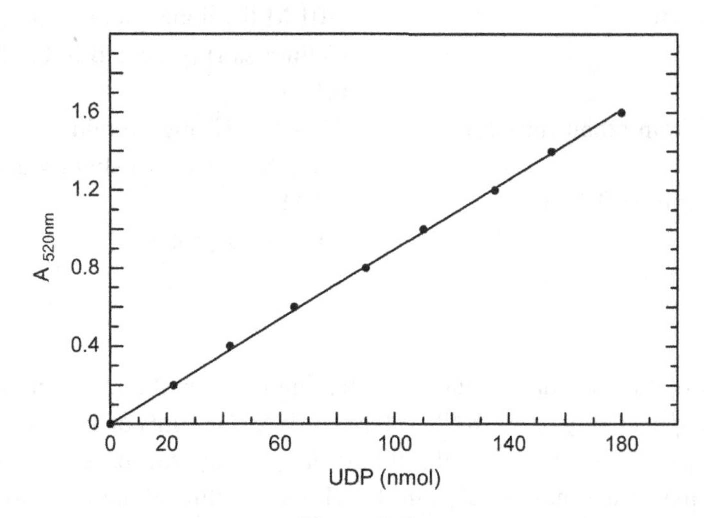

Figure 1.6
Standard curve for UDP determination with colorimetric method using 2,4-dinitrophenylhydrazine.

Comments

This method is used especially in studies of polysaccharide synthesis, and turbidity is avoided by the addition of alcohol. When ethanol can interfere by extracting pigments the reaction is stopped following this procedure:

> Add 500 μL of TCA, mix, and centrifuge. Transfer the supernatant to another tube. Add 150 μL of dinitrophenylhydrazine. Let stand for 5 min, and add 800 μL of 2.5 N NaOH. Mix and centrifuge. Determine absorbance at 520 nm.

This method can also be applied to determine ADP, GDP, TDP, and CDP.

1.16.2 Spectrophotometric Method Using Lactic Dehydrogenase

Principle

This method is based on the reduction of pyruvate by the action of lactate dehydrogenase in the presence of NADH and on determining NAD^+ formation (Kornberg, 1955). The enzymatic reactions to analyze UDP can be summarized as follows:

$$\text{UDP} + \text{Phosphoenolpyruvate} \xleftrightarrow{\text{Pyruvate kinase}} \text{UTP} + \text{Pyruvate}$$

$$\text{Pyruvate} + \text{NADH} + \text{H}^+ \xleftrightarrow{\text{Lactate dehydrogenase}} \text{Lactate} + \text{NAD}^+$$

Reagents

KCl	0.4 M
Phosphoenolpyruvate	0.01 M (cyclohexylammonium or sodium salt) (prepared in 0.4 M KCl)
$MgSO_4$	0.1 M
Pyruvate kinase (from rabbit muscle)	350–500 U · mg $protein^{-1}$
NADH	0.002 M (keep at a slightly alkaline pH)
Hepes–NaOH buffer (pH 7.4)	0.1 M
Lactate dehydrogenase (lactic acid dehydrogenase from rabbit muscle)	200 U · mg $protein^{-1}$

Procedure

To a 50-μL sample (or a standard solution containing between 2 and 50 nmol of UDP), add 25 μL of phosphoenolpyruvate-KCl, 100 μL of NADH, 1 mL of Hepes–NaOH buffer (pH 7.4), and 25 μL of an appropriate dilution of lactic dehydrogenase. Start the reaction by adding 25 μL of pyruvate kinase diluted in 0.1 M magnesium sulfate. Record the oxidation of NADH by the decrease in absorbance at 340 nm for 3 min at 30 s intervals.

To calculate UDP μmoles present in the sample, divide the absorbance decrease per min by 6.22, taking into account that the absorbance of 1 mL containing 1 μmol of NADH (the substrate that is being consumed) in a 1-cm path length cuvette at 340 nm is 6.22.

1.17 Determination of UDP-Glucose Using UDP-Glucose Dehydrogenase

Principle

The reaction catalyzed by the enzyme UDP-glucose dehydrogenase can be used to analyze UDP-glucose by measuring the formation of NADH spectrophotometrically at 340 nm (Strominger et al., 1957).

$$\text{UDP-glucose} + 2\,\text{NAD}^+ + \text{H}_2\text{O} \xrightarrow{\text{UDP-glucose dehydrogenase}} \text{UDP-glucuronic acid} + 2\,\text{NADH} + 2\,\text{H}^+$$

Reagents

Glycine buffer (pH 8.7)	1 M
NAD^+	0.05 M
UDP-glucose	0.01 M
UDP-glucose dehydrogenase (from bovine liver)	

Procedure

Add 10 μL of NAD^+, 50 μL glycine buffer (pH 8.7) and UDP-glucose dehydrogenase to a 10-μL sample (or a standard solution containing between 5 and 100 nmol of UDP-glucose). Make up to 500 μL with distilled water. After enzyme addition, incubate at 25–30°C, and read absorbance at 340 nm at 1-min intervals until no further reaction is detected. To calculate the amount of UDP-glucose (expressed in μmoles), divide the increase in absorbance at 340 nm by 12.44.

1.18 Determination of UDP-Glucose Using UDP-Glucose Pyrophosphorylase

Principle

The reaction catalyzed by the enzyme UDP-glucose phyrophosphorylase can be used to analyze UDP-glucose by measuring NADPH formation with the addition of phosphoglucomutase, glucose-6-phosphate dehydrogenase, and $NADP^+$, based on the following reactions:

$$\text{UDP-glucose} + \text{Inorganic pyrophosphate} \xleftrightarrow{\text{UDP-glucose phyrophosphorylase}} \text{UTP} + \text{Glucose-1-phosphate}$$

$$\text{Glucose-1-phosphate} \xleftrightarrow{\text{Phosphoglucomutase}} \text{Glucose-6-phosphate}$$

$$\text{Glucose-6-phosphate} + \text{NADP}^+ \xleftrightarrow{\text{Glucose-6-phosphate dehydrogenase}} \text{6-Phosphogluconate} + 2\text{NADPH}$$

The amount of the reduced product (NADPH), which is proportional to the quantity of glucose-1-phosphate, is estimated by the increase in absorbance at 340 nm (Pontis and Leloir, 1962).

Reagents

$Na_2P_2O_7$ (sodium pyrophosphate)	0.02 M
$MgCl_2$	0.1 M
Tris-HCl buffer (pH 7.8)	0.5 M
$NADP^+$	0.01 M
UDP-glucose pyrophosphorylase (from yeast)	$50\ U \cdot mg\ protein^{-1}$
Phosphoglucomutase (from yeast)	$400\ U \cdot mg\ protein^{-1}$
Glucose-6-phosphate dehydrogenase (from yeast)	$200\ U \cdot mg\ protein^{-1}$

Procedure

To a 50-μL sample (or a standard solution containing between 5 and 100 nmol of UDP-glucose) add 300 μL of Tris-HCl buffer (pH 7.8), 180 μL of $MgCl_2$, aliquots of the three enzymes (UDP-glucose pyrophosphorylase, phosphoglucomutase, and glucose-6-phosphate dehydrogenase appropriately diluted) and 20 μL of $NADP^+$. Make up to 970 μL final volume with distilled water. Start the reaction by adding 30 μL of pyrophosphate and incubate at 25–30°C in a 1-cm path cuvette. Read absorbance at 340 nm at 1-min intervals until no further reaction is detected. Run a control without pyrophosphate with every sample. To calculate the μmoles of UDP-glucose present in the sample, divide the increase in absorbance at 340 nm by 6.22.

Comments

Similar to the determination of UDP-glucose, ADP-glucose can also be quantified by using ADP-glucose pyrophosphorylase.

Further Reading and References

Albalasmeh, A.A., Berhe, A.A., Ghezzehei, T.A., 2013. A new method for rapid determination of carbohydrate and total carbon concentrations using UV spectrophotometry. Carbohydr. Polym. 97, 253–261.

Ashwell, G., 1957. Colorimetric analysis of sugars. In: Colowick, N., Kaplan, N. (Eds.), Methods Enzymol., vol. III. Academic Press, New York, pp. 73–105.

Brown, A.H., 1946. Determination of pentose in the presence of large quantities of glucose. Arch. Biochem. 11, 269–278.

Chifflet, S., Torriglia, A., Chisa, R., Tolosa, S., 1988. A method for determination of inorganic phosphate in the presence of labile organic phosphate and high concentration of protein: application to lend ATPases. Anal. Biochem. 168, 1–4.

Dahlqvist, A., 1961. Determination of maltase and isomaltase activities with a glucose-oxidase reagent. Biochem. J. 80, 547–551.

Dische, Z., 1949. Spectrophotometric method for the determination of free pentose and pentose in nucleotides. J. Biol. Chem. 181, 379–392.

Dreywood, R., 1946. Qualitative test for carbohydrate material. Ind. Eng. Chem. Anal. 18, 499.

Dubois, M., Guilles, K.A., Hamilton, J.K., Rebers, P.A., Smith, F., 1956. Colorimetric method for determination of sugars and related substances. Anal. Chem. 28, 350–356.

Elson, L.A., Morgan, W.T.J., 1933. Determination of N-acetylglucosamine acid N-acetylcondrosamine. Biochem. J. 27, 1824–1828.

Fiske, C.H., Subbarow, Y., 1925. The colourimetric determination of phosphorus. J. Biol. Chem. 66, 375–381.

Jones, M.G., Outlaw, W.H., Lowry, O.H., 1977. Enzymic assay of 10^{-7} to 10^{-14} moles of sucrose in plant tissues. Plant Physiol. 60, 379–383.

Kornberg, A., 1955. Lactic dehydrogenase of muscle. In: Colowick, N., Kaplan, N. (Eds.), Methods. Enzymol., vol. I. Academic Press, New York, pp. 441–443.

Leloir, L.F., Cardini, C.E., 1957. Characterization of phosphorus compounds by acid lability. In: Colowick, N., Kaplan, N. (Eds.), Methods. Enzymol., vol. III. Academic Press, New York, pp. 840–850.

Lowry, O.H., Carter, J.G., 1974. Stabilizing the alkali-generated fluorescent derivatives of NAD and NADP. Anal. Biochem. 59, 639–642.

Morgan, W.T.J., Elson, L.A., 1934. A colourimetric method for the determination of N-acetylglucosamine and N-acetylchrondrosamine. Biochem. J. 28, 988–995.

Nelson, N., 1944. A photometric adaptation of the Somogyi method for the determination of glucose. J. Biol. Chem. 153, 375–380.

Okuda, J., Miwa, I., Maeda, K., Tokui, K., 1977. Rapid and sensitive, colourimetric determination of the anomers of D-glucose with D-glucose oxidase, peroxidase, and mutarotase. Carbohy. Res. 58, 267–270.

Percheron, F., 1962. Dosage colourimetrique du fructose et des fructofuranosides par l'acide thiobarbiturique. C.R. Hebd. Séances Acad. Sci. 255, 2521–2522.

Pontis, H.G., Leloir, L.F., 1962. Measurement of UDP-enzymes systems. In: Glick, D. (Ed.), Methods of Biochemical Analysis. Interscience Publishers, J. Wiley and Sons, New York, London, pp. 107–136.

Reissig, J.L., Storminger, J.L., Leloir, L.F., 1955. A modified colourimetric method for the estimation of N-acetylamino sugars. J. Biol. Chem. 217, 959–966.

Somogyi, M., 1952. Determination of reducing sugars. J. Biol. Chem. 195, 19–23.

Strominger, J.L., Maxwell, E.S., Axelrod, J., Kalckar, H.M., 1957. Enzymatic formation of uridine diphosphoglucuronic acid. J. Biol. Chem. 224, 79–81.

Sturgeon, R.J., 1980. Enzymatic determination of D-glucose, D-fructose, and D-mannose. Methods Carbohydr. Chem. 8, 135–137.

Zhou, M., Diwu, Z., Panchuk-Voloshina, N., Haugland, R.P., 1997. A stable nonfluorescent derivative of resorufin for the fluorometric determination of trace hydrogen peroxide: applications in detecting the activity of phagocyte NADPH oxidase and other oxidases. Anal. Biochem. 253, 162–168.

PART II

Purification and Analysis of Proteins and Carbohydrates

CHAPTER 2

Preparation of Protein Extracts

Chapter Outline

2.1 Introduction

It is the purpose of this chapter to outline the principles of the techniques generally used to prepare enzymically active extracts from tissues of photosynthetic organisms, such as plants, unicellular green algae, and cyanobacteria. In particular, the content will be based on studies of enzymes involved in carbohydrate metabolism.

The objective of plant protein extraction and purification has changed in the last decades. Before the genomic era, the starting question was what plant species and/or tissue was most favorable for purification of large amounts of the enzyme that was the object of biochemical, structural, or other studies. Often purification to homogeneity of an active protein from a plant material was the most common limitation. At present, with the availability of sequenced genomes and genetic engineering methodologies, most studies are performed from recombinant proteins produced in heterologous systems. However, protein studies of native enzymes from plant, algae, or cyanobacteria are still needed (eg, to assay an enzyme in a crude extract in physiological studies, or to corroborate expression analysis or functional identification of genes). In these cases, the source material is not a choice and the preparation of active extracts is mandatory.

Methods for Analysis of Carbohydrate Metabolism in Photosynthetic Organisms.
DOI: http://dx.doi.org/10.1016/B978-0-12-803396-8.00002-8

There is no universal method for extracting proteins from plant cells since they contain a wide range of proteins that vary greatly in their properties and stability and, therefore, may need specific conditions in the cell disruption process. Similarly, there is no universal buffer to suspend a tissue for homogenization. It may be useful to consult the literature about the mediums used to extract similar proteins to those to be studied; however, it should be kept in mind that in most cases the method of choice is reached by trial and error and can always be optimized. Anyway, the extraction procedures depend on the nature of the biological material (tissue, organ, or cell) and the characteristics of the enzyme being studied (protein stability and/or activity lability, redox status, solubility, etc.). Some of the specific problems that must be considered when developing protocols for protein extraction from plant tissues or photosynthetic microorganisms are the presence of rigid cell walls (containing, eg, cellulosic-type compounds or sporopollenin as it occurs in plants or in some microalgae), which must be sheared to release the cell content, and specific contaminating compounds that may affect protein integrity and/or stability (eg, phenolics and polyphenolics, organic acids, and a range of proteinases). Sometimes it is possible to choose the biological material that does not have such interferences. However, in specific studies, this is not possible and it is necessary to optimize the extraction conditions and/or to find the ways to remove or inactivate the undesirable contaminants. Also, when the enzyme of interest is organellar, it should be extracted from isolated organelles or from an organelle-enriched cytosolic extract. These specific considerations will not be discussed here.

General and useful methods for extracting active proteins from plants, algae, and cyanobacteria and current specific protocols used in studies of enzymes involved in carbohydrate metabolism (most of them developed in our laboratory) will be described below.

2.2 Extraction of Proteins

2.2.1 Source Material

Since the objective is to obtain active proteins, special considerations must be taken starting from the biological material. Although the ideal source is freshly harvested tissues, organs, or cells, other various types of material are employed for enzyme extraction, such as frozen tissues or cells (at −80°C, or at ~−196°C in liquid nitrogen), or freeze-dried or acetone powder materials. However, when the purpose is the purification of organelle activities, freshly harvested biological material is mandatory. For example, for preparation of chloroplastic enzymes, the most freshly harvested leaves are said to give the most consistent results.

Nevertheless, in most cases, the wet material can be frozen under liquid nitrogen immediately after harvesting, and stored at −80°C. The storage temperature is critical to preserve protein integrity and enzyme activity. Freezing biological tissues to temperatures above the eutectic point of some salts (such as potassium chloride and phosphates, natural cell components) can cause unexpected pH changes and protein denaturation. Also, storage at −20°C is not enough for preventing proteolytic digestion, since at that temperature hydrolytic enzymes can be released and act on their substrates. After extraction, enzymes of interest may be still active but it is possible that they have been damaged.

Alternatively, leaves and some other tissues (such as seeds, seedlings, etc.) can be treated with acetone at −20°C (that displaces the water) in a blender. The resulting acetone powder can be kept active, after being dried and stored at low temperature (−20°C). Later, when it is necessary, the powder can be extracted in an appropriate medium. This preparation is almost free from chlorophyll, lipids, and resinous compounds.

Also, freeze-drying (method of removing water by sublimation of ice crystals from frozen material) of cell suspensions or plant tissues is a gentle technique that can be used in many cases. Proteins can be extracted directly from the powder obtained in the presence of an extraction solution.

2.2.2 Homogenization Buffer Composition

The extraction of a soluble protein from a plant tissue varies considerably according to the source material. For example, to make extracts from wheat germ, grinded seed powder, or lyophilized material, it is sufficient to make a suspension in a suitable extraction solution for a certain time. In contrast, the first step for preparation of protein extracts from most tissues, organs, or cells is the disruption in an adequate homogenization buffer to release cell contents, taking care to protect proteins from damage. A volume of extraction solution (as minimal as possible) is added to the biological material. In the case of plants, this volume is less than $1\ mL \cdot g^{-1}$ of fresh tissue, and for microorganism homogenization, at least $2\ mL \cdot g^{-1}$. The protein solvents usually contain several solutes to improve stability, keep the protein in solution, prevent the microorganism growth, and/or reduce the freezing point during storage. Importantly, the solution is always buffered to give stability to the proteins. Usually, in most cases, the pH should be measured and adjusted during the extraction.

A typical homogenization solution for plant tissues includes high concentrations of reductant, polyphenol inactivators, and protease inhibitors. To define the composition and pH of the buffer solution it is important to stress that the proteins are least soluble at their isoelectric point. Thus, the extraction should be carried out at a pH far from it, but it should be such that the enzyme is stable. In general, extracts are made at a pH of about 7.0 ± 1.0.

Salts (eg, KCl, NaCl, $(NH_4)SO_4$) and organic compounds (eg, sucrose and glycerol) are added to the extraction medium on empirical bases, considering that the enzyme of interest is soluble while the others are not, and thereby promoting purification from the beginning. Sometimes the solubility and stability of enzymes are enhanced by the addition of substrates, ligands, or effectors (eg, Mg^{2+}, ATP, or phosphate), detergents (such as deoxychlolate), and/or surfactants (such as Triton X-100). The addition of reducing agents (such as β-mercaptoethanol (at 5–14 mM), dithiothreitol (DTT, at 2–5 mM), dithioerythritol, reduced glutathione, or cysteine, or also a thiol-protease inhibitor (at 10–30 mM)) to the extraction medium helps to maintain essential sulfhydryls in the reduced state. It is also convenient to include protease inhibitors like phenylmethyl sulfonyl fluoride (PMSF, at 0.5–1 mM), N-tosyl-L-phenylalanine choro-methyl ketone, benzamidine.HCl (or p-aminobenzamidine.2HCl, at 1 mM), ϵ-amino-n-caproic acid (at 5 mM), chymostatin (at 10.5 $\mu g \cdot mL^{-1}$), leupetine or antipain (at 1–5 $\mu g \cdot mL^{-1}$), and aspartate protease inhibitors (such as pestatin or diazoacetylnorleucine methyl ester). Alternatively, bovine albumin can be added to the medium, acting as an artificial substrate for proteases that reduces proteolytic degradation of the enzyme of interest, and can act as a stabilizer (ie, protecting proteins from denaturation in diluted solutions).

Additionally, the removal of tannins and other phenolic compounds from protein extracts can be done by filtration after addition to the medium of insoluble polyvinylpolypyrrolidone (PVPP, 1.5% (w/v)) or a polymer as polyclar, or by addition of soluble polyvinylpyrrolidone (PVP, Mr 40,000, 2–4%). Also, the maintaining of a strong reducing environment (eg, by adding DTT, β-mercaptoethanol, sodium diethyl dithiocarbamate (at 20 mM), sodium metabisulfite (at 10–20 mM), or buffers containing borate or sodium tetraborate (at 0.2 M)) counteracts the effect of phenol oxidases. Chelating agents, such as ethylene diamine tetraacetic acid (EDTA, between 1 to 2 mM) or ethylene glycol tetraacetic acid (EGTA), may be added to the extraction medium to remove calcium and magnesium ions, or heavy metals that can interfere with activity. Sodium fluoride (at 20 mM) is usually included in the extraction solution to prevent phosphatase activity.

2.2.3 Breakdown of Biological Material (Cells and Tissues Disruption)

In contrast to animal cells that are easy to disrupt, the first and most important problem to overcome for extracting active proteins from plants and photosynthetic microorganisms is to actually break up the cells to release the active enzymes. This operation, usually carried out at 0–4°C, should be done rapidly to minimize the exposure of proteins to damaging compounds or undesirable activities.

The use of a blender or an homogenizer, grinding together with an abrasive (usually under liquid nitrogen), ultrasonic disintegration (using a sonicator), disruption by pressure release (eg, using a French Press), and solvent solubilization of cell wall material are among the most common procedures used. The choice depends on the work scale and type of material. Anyway, it is advisable to test different methods on a small scale in order to assess the amount and quality of the protein of interest released into the extract.

The use of a blender is a simple and ideal method for disrupting plant tissue by shear forces. The plant material is cut into small pieces and blended in the presence of the extraction buffer, for about 1 min and then centrifuged to remove debris. On the other hand, disruption in a Potter–Elvehjem homogenizer (cylindrical glass or hard polymer pestle that rotates in a close-fitting tube), which is generally used for animal tissues, can be used in special cases such as young leaves or seedlings. Following a similar principle, homogenates from small amounts of tissue or cells (10–100 mg) can be obtained using eppendorf tubes and a rotating conic-finished glass stick.

Grinding of frozen plant material to a fine powder using a mortar and pestle in the presence of abrasives (such as acid-washed sand, alumina, or glass powder) and a small amount of extraction buffer, under liquid nitrogen is, in general, a useful method for plant, algae, and cyanobacterial protein extract preparations.

Sonication (ultrasonic homogeneization) is a good option for the disruption of suspended cells (c.10–100 mL). A sonicator probe is lowered into the suspension which is subjected to high frequency sound waves for short periods (30–60 s) two or three times. These waves cause disruption of cells by shear force and cavitation, which refers to areas where there is alternate compression and rarefaction. Because considerable heat is generated during the procedure, special care to keep the extract at low temperature should be taken.

The use of a press (such as a French Press) is an excellent means for disrupting cells of green algae or cyanobacteria. The principle of the technique is simple and involves placing a prechilled (at 4°C) cell suspension (normally ratios of 1:1 to 1:4, cell wet weight:buffer volume) into the center hole of a column of stainless steel. The suspension is subjected to very high pressure (7000–10,000 psi) and forced by a piston-type pump through a small orifice, generating instantaneous disruption of the cells. This procedure is usually repeated twice or more times, to lyse all the cells under carefully controlled conditions. A similar alternative method is the usage of the Hughes Press, in which the cells are forced through the orifice as a frozen paste, often mixed with an abrasive, aiding the disruption of the cell walls.

2.2.4 Cell Permeabilization

Determination of enzyme activity in some unicellular algae (such as *Chlorella vulgaris*, *Scenedesmus obliquus*, *Prototheca zopfii*, and *Chlamydomonas reinhardii*) has often been a problem because drastic treatments are required to break the cells due to their rigid and undigestable cell walls. An alternative approach shown to be successful for the analysis of metabolic processes in algae and cyanobacteria is the permeabilization of cells with toluene (eg, 2%, volume of toluene/volume of cell suspension) after mixing with a vortex twice for 1–3 min. Advantages of this method are: (1) enzymes can be determined under conditions resembling more the in vivo situation; (2) it is a rapid and simple method; and (3) very low amounts of cells are required. For example, for enzyme activity assays in cyanobacteria and algae, about 80 mg of cells per mL of extraction buffer were used.

2.3 Laboratory Procedures for Protein Extraction

2.3.1 Preparation of Acetone Powder from Wheat Leaves (Calderón and Pontis, 1985)

Materials

Pure acetone (at −20°C)
Blender (Waring Blendor) with 1.5–2 L glass
Büchner funnel (20–30 cm diameter) and a 2-L Kitasato flask
Vacuum pump (or water jet pump)
Straight and curved spatulas and steel clamps
15 cm-diameter crystallizer
Filter paper
Desiccator with pure sulfuric acid (as desiccant compound).

Procedure

Carry out all operations at 2–4°C, if possible in a cold room. Cut leaves from 10-day-old wheat seedlings into segments (c.20–30 g) and transfer to the blender glass. Add 1 L of acetone and blend for 30 s–1 min at maximum speed. Prepare in advance the Büchner funnel with the filter paper, and the Kitasato flask connected to a vacuum pump (or a water jet pump). Filter the acetone suspension, taking care to not pass air. Stop the vacuum when the acetone has been totally filtered. Carefully lift the filter paper with the resultant powder with a clamp; transfer the powder gently with a dry straight spatula to the glass blender containing another liter of acetone. Repeat the homogenization and the filtration through the Büchner funnel twice. In the last filtration, press the powder against the filter paper using a curved spatula, taking care to not pass air. After filtration, transfer the resulting powder to a 15 cm-diameter crystallizer. Cover with a filter paper and fasten with a rubber band to the crystallizer. Make two or three holes in the filter paper with the spatula tip. Place the crystallizer inside a desiccator containing sulfuric acid (p.a. grade) as a drying agent. Apply a vacuum gently to avoid letting the acid "boil." Leave it in the desiccator overnight to complete the removal of the residual acetone and store the powder at −20°C until use.

To prepare a protein extract, weigh an amount of powder, and resuspend it in the selected extraction buffer for 1 h, with intermittent stirring at 4°C. Centrifuge the resulting homogenate at 20,000 × g for 15 min. The supernatant can be used for enzyme activity determinations.

2.3.2 Preparation of Plant Tissue Extracts

Homogenates from plant tissues were successfully carried out from *Arabidopsis*, wheat, rice, and spinach leaves, seedlings, or roots to assay carbohydrate metabolism enzymes. Basically crude extracts were prepared as described below.

2.3.2.1 Preparation of leaf or seedling extracts

Material

Wheat leaves (fresh or frozen at −80°C)
Extraction buffer: 100 mM Hepes–NaOH (pH 7.5), 1 mM EDTA, 20 mM $MgCl_2$, 20% (v/v) glycerol, 0.01% (v/v) Tritón X-100, 20 mM β-mercaptoethanol, and 1 mM phenylmethylsulfonyl fluoride.

Note: Add β-mercaptoethanol and phenylmethylsulfonyl fluoride to the extraction buffer just before use.

Procedure

Transfer the leaves (c.100 g fresh weight) into a precooled mortar (at −20°C) containing liquid nitrogen. Grind the material with a pestle until a fine powder is obtained (prevent thawing by adding liquid nitrogen, if necessary). Add 1.5 mL of cold extraction buffer per gram of fresh weight. Allow extraction during 10–15 min at 0–4°C with occasional stirring with a glass rod. Filtrate the homogenate through a nylon tissue (50 μm-mesh). Centrifuge the filtrate at 30,000 × g for 20 min. Desalt the supernatant (see Section 2.4) before enzyme activity determination.

Note: A similar procedure was used for protein extraction from c.200–500 mg fresh weight of wheat, *Arabidopsis*, and rice leaves or wheat roots. In the last case, the extraction buffer contained 100 mM Tris-HCl (pH 7.2) and 20 mM β-mercaptoethanol for 1 h at 0°C and the extracts were centrifuged in sinterglass filter units at 2000 × g at 4°C.

2.3.2.2 Preparation of protein extracts from spinach and Arabidopsis chloroplasts

This method is an adaptation of that described by Mourioux and Douce (1981) and Schuler and Zielinski (1989).

Materials

Percoll

Grinding buffer: 50 mM Hepes–NaOH (pH 7.5) containing 2 mM EDTA, 400 mM mannitol, 1 mM $MgCl_2$, 0.2% β-mercaptoethanol, and 0.2% (w/v) bovine serum albumin.

Protein extraction buffer: 50 mM Hepes–NaOH (pH 7.5), 1 mM EDTA, 20 mM $MgCl_2$, 20% (v/v) glycerol, 0.01% Triton X-100, 20 mM β-mercaptoethanol, and 1 mM phenylmethylsulfonyl fluoride.

Procedure

(A) Preparation of intact chloroplasts

Prepare the Percoll gradient solution in 50-mL centrifuge tubes, by pipetting 17.5 mL of Percoll and 17.5 mL grinding buffer and centrifuging at 22,000 × g, for 45 min at 4°C (brake off).

Immediately after the initiation of the centrifugation, cut the leaves (c.50 g) into short fragments, and cool them in ice water for 15 min. Transfer the leaf segments to a mortar (precooled at 0°C). Add 5 mL of grinding buffer per 1 g of tissue. Homogenize gently with a pestle. Filter the resulting suspension through a nylon cloth (50 μm-mesh). Centrifuge the filtrate at 2500 × g at 4°C for 4 min. Discard the supernatant. Wash the precipitate three times with the extraction solution. Finally, gently resuspend the pellet in 6 mL of the same solution.

Load carefully the suspension on top of a preformed Percoll gradient, avoiding disturbing the gradient. Centrifuge at 9000 × g at 4°C for 15 min without the brake on. Carefully remove the gradients from the rotor. Intact chloroplasts (lower band) will be separated from other cell components (upper band containing broken chloroplasts, and other cell remnants).

Carefully remove each band with a Pasteur pipette. Place the chloroplast fraction into a centrifuge tube and wash three times (add 20 mL of grinding buffer each time) gently invert the tube capped with parafilm and centrifuge at 2500 × g for 4 min at 4°C. Check the integrity of the plastids by optical microscopy. Use chloroplast for preparation of protein extracts.

(B) Protein extract preparation

Suspend the chloroplast fraction in extraction buffer (ratio: 1 g fresh weight:1.5 mL of buffer). Homogenize the chloroplast paste in a mortar with a pestle under liquid nitrogen and glass beads (200 μm diameter). Centrifuge the extract at 30,000 × g for 20 min at 4°C. Desalt the supernatant through a Sephadex G-50 column equilibrated in the protein extraction buffer. Use the eluate for enzyme activity determination.

2.3.3 Preparation of Extracts from Photosynthetic Microorganisms

Representative procedures for protein extraction from model unicellular algae (*Chlorella vulgaris*, *Chlamydomonas reinhardtii*, and *Ostreococcus tauri*), unicellular cyanobacteria (*Synechocystis* sp. PCC 6803, *Synechococcus* sp. PCC 7002, *Microsystis aeruginosa* PCC 7806), filamentous cyanobacteria (*Anabaena* (*Nostoc*) sp. PCC 7120) cyanobacteria, and a single-celled flagellate photosynhtetic protist (*Euglena gracilis*) will be described below.

2.3.3.1 Chlorella vulgaris

Materials

Chlorella vulgaris (Beijerinck strain 11468) cultures (500 mL)
Washing buffer: 25 mM Tris-HCl (pH between 7.0 to 8.0) containing 1 mM EDTA and 5 mM β-mercaptoethanol
Extraction buffer: 100 mM Hepes-NaOH (pH 7.5) containing 20 mM β-mercaptoethanol, 2 mM EDTA, 2% ethylenglycol, and 0.5 mM phenylmethylsulfonyl fluoride
Glass powder

Procedure

Carry out all operations at 2–4°C. Collect the cells by centrifugation at 3000 × g for 5 min and wash twice with the washing buffer. Resuspend packed cells in five to eight times their volume with extraction buffer. Cells can be broken either by sonication (c.0.1–1 g of fresh weight) at 40–100 W (eg, 3 pulses of 10 min) in the presence of glass powder (keeping refrigerated at 10–12°C), or by passage through a French press (c.3–15 g of fresh weight) at 25,000 psi. Check disintegration of cells by light microscope observation. After removing cell debris by centrifugation (30,000 × g for 30 min) and desalting, the extract can be used for enzyme activity determination. Eventually, proteins can be concentrated with solid ammonium sulfate bringing the supernatant to 70% saturation, while the pH is kept at 7.0 by addition of ammonium hydroxide, or with an ultrafiltration concentration system. Desalt the protein extract before enzyme activity determination.

Comments

Although sonication and French press disruption are capable of disintegrating all cells, usually higher enzyme activities can be obtained when c.80% of total protein is released to the medium. A similar procedure was used in the study of the sugar metabolism enzymes of *Scenedesmus obliquus* (strain 11457) and *Prototheca zopfii*.

2.3.3.2 *Chlamydomonas reinhardtii*

Materials

Chlamydomonas reinhardtii (strain CC124) cultures (1 L)
Washing buffer: 10 mM Tris-HCl (pH 8.0) containing 1 mM EDTA (pH 8.0)
Extraction buffer: 100 mM Hepes-NaOH (pH 7.5) containing 20 mM β-mercaptoethanol, 2 mM EDTA (pH 7.0), 20 mM $MgCl_2$, 0.5 mM phenylmethylsulfonyl fluoride, 20% glycerol, and 2% ethylenglycol
Glass powder

Procedure

Carry out all operations at 2–4°C. Collect the cells by centrifugation at 3000 × g for 5 min, wash the cell pellet with washing buffer, precipitate the cells by centrifugation at 3000 × g for 5 min, and resuspend them in 20 mL of extraction buffer. Disruption of cells can be performed by two cycles of slow freezing to −80°C followed by thawing to room temperature, or with a French press at 700 psi (high). Centrifuge the homogenate at 30,000 × g for 20 min. Measure enzyme activity in the supernatant and/or proceed to further purification.

Comments

The *Chlamydomonas* cell structure can be also disrupted by brief exposure to sonication (a total of 30–60 s at 4°C at 30–40 W). Acetone powders for enzyme extraction can also be easily prepared with *Chlamydomonas* cells.

A procedure similar to those described for unicellular algae can be used to prepare protein extracts from *Euglena gracilis* (a single-celled flagellate protist). The extraction buffer consists of 50 mM Hepes-NaOH (pH 7.5) containing 20 mM β-mercaptoethanol, $MgCl_2$ 20 mM, 2 mM EDTA, 2% ethylenglycol, 0.5 mM phenylmethylsulfonyl fluoride, and glycerol 20%, and breaking cells by sonication (0.1–1 g fresh weight of cells, 3 pulses of 40 W for 10 s) or with a French press (3–15 g fresh weight of cells at 3000 psi).

2.3.3.3 *Ostreococcus tauri*

Materials

Ostrococcus tauri (strain 0TTH0595) cultures (600 mL)
Washing buffer: 25 mM Tris-HCl (pH 7.5) containing 1 mM EDTA and 5 mM β-mercaptoethanol
Extraction buffer: 50 mM Tris-HCl (pH 7.5) containing 1 mM EDTA, 100 mM NaCl, 20 mM β-mercaptoethanol, and 1 mM phenylmethylsulfonyl fluoride
Glass powder

Procedure

Carry out all operations at 2–4°C. Collect the cells by centrifugation at 3000 × g for 5 min and wash twice with the washing buffer. Resuspend packed cells in five volumes of extraction buffer. Conduct three cycles of freezing in liquid nitrogen/thawing. Then, break the cells in the presence of glass powder either with a Potter–Elvehjem homogenizer with conic-finished glass stick with Teflon tip or by sonication (using a 2 mm-tip probe, 10 cycles of 10 s at 40 W, with 20 s pauses). Centrifuge 30,000 × g for 30 min to remove cell debris and desalt the supernatant through Sephadex G-50 equilibrated with the extraction buffer.

2.3.3.4 Cyanobacteria

Materials

Anabaena (also named *Nostoc*) sp. PCC 7120 cultures
Washing buffer: 25 mM Hepes–NaOH (pH 7.5) containing 2 mM EDTA (pH 7.0) and 5 mM β-mercaptoethanol
Extraction buffer: 100 mM Hepes–NaOH (pH 7.5) containing 2 mM EDTA (pH 7.0), 2% (v/v) ethylene glycol, 20% (v/v) glycerol, 20 mM $MgCl_2$, 20 mM β-mercaptoethanol, and 1 mM phenylmethylsulfonyl fluoride
Small glass beads (<200 μm in diameter) or glass powder

Procedure A

Collect cells (from c.150–300 mL) by centrifugation at 3000 × g for 10 min. Wash the pellet twice by addition of washing buffer and centrifuging each time at 3000 × g for 10 min.

Resuspend the packaged cells in the extraction buffer (2 mL of buffer per gram of fresh weight) and distribute in five eppendorf tubes. Add glass beads (or glass powder) to the cell paste and freeze in liquid nitrogen. Submit the cell paste contained in each tube to five cycles of freezing and disintegration with a Teflon tipped glass rod (or with a conic-finished frosted glass stick), precooled with liquid nitrogen, placed in a vertical laboratory stirrer for 30 s each cycle. Centrifuge the extract 30,000 × g for 30 min to remove cell debris. Take the supernatant with a Pasteur pipette and place the liquid in a cooled tube. Desalt the clarified protein extract through a Sephadex G-50 bed (see Section 2.4).

Procedure B

Collect cells by centrifugation and add two times their volume of washing buffer. Centrifuge and weigh the pellet (c.3–8 g of cells). Homogenize the cell paste in the presence of glass beads and extraction buffer (2 mL of buffer per gram of fresh weight) in a −20°C-precooled mortar using a pestle, under liquid nitrogen. Sonicate the extract for four cycles of 30 s at 4°C, at 40 W. Centrifuge at 30,000 × g for 30 min to remove debris and then, centrifuge the supernatant at 100,000 × g for 1 h. Desalt the supernatant through a disposable (or reusable) desalting column or as described in Section 2.4 before the enzyme activity assay.

Comments

Alternatively, protein extraction can be carried out from −80°C-stored cells that have been harvested, washed, and weighed before storage. Usually, enzyme activities are lower when the extract is prepared from frozen cells than from fresh cells.

Similar extraction procedures can be applied for protein extraction from *Syechocystis* sp. PCC 6803, *Synechococcus* sp. PCC 7002, or *Microsyctis aeruginosa* PCC 7806.

2.4 Protein Extract Desalting

The clean up of proteins from low molecular weight molecules (such as salts, tannins, and other phenolic compounds) that can interfere with methods of enzyme activity assays can be carried out by passage through a resin (Sephadex G-25 or G-50) packaged in small columns. These columns can be purchased from a supplier (as filtration desalting columns) or prepared in the laboratory.

Materials

5-mL syringes (or eppendorf tubes)
Extraction buffer (according to the biological material to be extracted)
Distilled water
Equilibrated resin (Sephadex G-25 fine or G-50) with the same extraction buffer

Procedure

Up to 0.8 mL of extract to be desalted, use a 5-mL syringe with three filter paper circles in the bottom or an eppendorf tube with a hole in the bottom with a glass wool bed (c.1 mL of resin for 0.3–0.5 mL of extract). Load the syringe (or the eppendorf tube) with a volume of resin of approximately 2:1 (height:diameter ratio). Centrifuge the syringe resting on a centrifuge tube at 3000 × g for 10 min at 4°C (these parameters should be exactly repeated in order to have reproducible results). Load an aliquot of the extract on the flat top of the resin. Elute the protein by centrifugation at 3000 × g at 4°C for 10 min, collecting the eluate in a new centrifuge tube. The extract is ready for enzyme activity determination.

Resin regeneration

Carry out the procedure in batches, placing the resin (G-25 or G-50) in a beaker. Wash the resin twice with distilled water (10:1, volume of water:volume of resin), agitating with a glass rod or with a magnetic stirrer. Let the resin decant each time and discard the supernatant carefully. Repeat the above procedure twice adding NaCl 0.5 M (10:1, volume of salt:volume of resin). Wash the resin four times with distilled water (10:1, volume of water:volume of resin) under agitation. Add sodium azide 0.02% to the resin suspended in water. Store the regenerated resin at 4°C. Before use, the resin should be thoroughly washed with distilled water (five to six times its volume).

Comments

Protein content should be measured (eg, with the Bradford reagent) in the extract before and after desalting to check that there is no loss of protein by retention in the resin. To optimize the method, the volume of extract loaded should be similar to the volume eluted after centrifugation.

Further Reading and References

Calderón, P., Pontis, H.G., 1985. Increase of sucrose synthase activity in wheat plants after a chilling shock. Plant Sci. 42, 173–176.

Cavalcanti, E.D., Maciel, F.M., Villeneuve, P., Machado, O.L., Freire, D.M., 2007. Acetone powder from dormant seeds of *Ricinus communis* L. Lipase activity and presence of toxic and allergenic compounds. Appl. Biochem. Biotechnol. 137–140 (1–12), 57–65.

Crespi, M.D., Zabaleta, E.J., Pontis, H.G., Salerno, G.L., 1991. Sucrose synthase expression during cold acclimation in wheat. Plant Physiol. 96, 887–891.

Cumino, A., Ekeroth, C., Salerno, G.L., 2001. Sucrose-phosphate phosphatase from *Anabaena* sp. strain PCC 7120: Isolation of the protein and gene revealed significant structural differences from the higher-plant enzyme. Planta. 214, 250–256.

Deutscher, M.P., 1990. Maintaining protein stability. In: Deutscher, M.P. (Ed.), Methods Enzymology, vol. 182. Academic Press, San Diego, CA, pp. 83–89.

Duran, R., Pontis, H.G., 1977. Sucrose metabolism in green algae I. The presence of sucrose synthetase and sucrose phosphate synthetase. Mol. Cell. Biochem. 16, 149–152.

Felix, H., 1982. Permeabilized cells. Anal. Biochem. 120, 211–234.

Fiol, D.F., Salerno, G.L., 2005. Trehalose synthesis in *Euglena gracilis* (Euglenophyceae) occurs through an enzyme complex. J. Phycol. 41, 812–818.

Foresi, N., Correa-Aragunde, N., Parisi, G., Caló, G., Salerno, G., Lamattina, L., 2010. Characterization of a nitric oxide synthase from the plant kingdom: NO generation from the green alga *Ostreococcus tauri* is light irradiance and growth phase dependent. Plant Cell. 22, 3816–3830.

Gegenheimer, P., 1990. Preparation of extracts from plants. In: Deutscher, M.P. (Ed.), Methods Enzymology, vol. 182. Academic Press, San Diego, CA, pp. 174–193.

Harris, E.H., 2009. The chlamydomonas source book. In: second ed. Harris, E.H., Stern, D.B., Witman, G.B. (Eds.), Introduction to Chlamydomonas and Its Laboratory Use, vol. 1. Elsevier Inc, Amsterdam.

Kolman, M.A., Torres, L.L., Martín, M.L., Salerno, G.L., 2012. Sucrose synthase in unicellular cyanobacteria and its relationship with salt and hypoxic stress. Planta. 235, 955–964.

Martin, M.L., Zabaleta, E.J., Lechner, L., Salerno, G.L., 2013. A mitochondrial alkaline/neutral invertase isoform (A/N-InvC) functions in developmental energy-demanding processes in *Arabidopsis*. Planta. 237, 813–822.

Martínez-Noël, G.M., Cumino, A.C., Kolman, M.A., Salerno, G.L., 2013. First evidence of sucrose biosynthesis by single cyanobacterial bimodular proteins. FEBS Lett. 587, 1669–1674.

Mourioux, G., Douce, R., 1981. Slow passive diffusion of orthophosphate between intact isolated chloroplasts and suspending medium. Plant Physiol. 67, 470–473.

Perez-Cenci, M., Salerno, G.L., 2014. Functional characterization of *Synechococcus* amylosucrase and fructokinase encoding genes discovers two novel actors on the stage of cyanobacterial sucrose metabolism. Plant Sci. 224, 95–102.

Porchia, A.C., Salerno, G.L., 1996. Sucrose biosynthesis in a prokaryotic organism: presence of two sucrose-phosphate synthases in *Anabaena* with remarkable differences compared with the plant enzymes. Proc. Natl. Acad. Sci. USA 93, 13600–13604.

Salerno, G.L., 1985a. Occurrence of sucrose and sucrose metabolizing enzymes in achlorophyllous algae. Plant Sci. 42, 5–8.

Salerno, G.L., 1985b. Measurement of enzymes related to sucrose metabolism in permeabilized *Chlorella vulgaris* cells. Physiol. Plant. 64, 259–264.

Salerno, G.L., Pontis, H.G., 1989. Raffinose synthesis in *Chlorella vulgaris* cultures after a cold shock. Plant Physiol. 89, 648–651.

Santoiani, C.S., Tognetti, J.A., Pontis, H.G., Salerno, G.L., 1993. Sucrose and fructan metabolism in wheat roots at chilling temperatures. Physiol. Plant 87, 84–88.

Schuler, M.A., Zielinski, R.E., 1989. Preparation of intact chloroplasts from pea. In: Methods in Plant Molecular Biology, Academic Press, San Diego, pp. 39–47.

Scopes, R.K., 1978. Techniques for protein purification. In: Kornberg, H.L. (Ed.), Techniques in Protein and Enzyme Biochemistry, Techniques in the Life Sciences: Biochemistry, vol. B1. Elsevier, The Netherlands, pp. 1–42.

Stoll, V.S., Blanchard, J.S., 1990. Buffers: principles and practice. In: Deutscher, M.P. (Ed.), Methods Enzymology, vol. 182. Academic Press, San Diego, CA, pp. 24–37.

Vargas, W.A., Pontis, H.G., Salerno, G.L., 2008. New insights on sucrose metabolism: evidence for an active A/N-Inv in chloroplasts uncovers a novel component of the intracellular carbon trafficking. Planta. 227, 795–807.

CHAPTER 3

Protein and Carbohydrate Separation and Purification

Chapter Outline

3.1 Introduction

The purification of proteins and carbohydrates from a biological source is currently required in biochemical and physiological studies and/or biotechnological projects. Two different approaches can be employed to obtain proteins or carbohydrates: (1) for identification and/or quantitation (analytical scale, small amounts are required) and (2) for performing further studies (large scale) (eg, in the case of proteins, for determination of enzyme biochemical properties, structure, etc., and in the case of carbohydrates, to study their chemistry or biological properties). Nonetheless, separation and purification of proteins or carbohydrates are not always required (eg, enzyme assays in crude extracts, or in situ or in permeabilized tissues or cells, or when the carbohydrate molecule exhibits some unique property that allows direct quantitation). When the separation of the cell components is necessary, a suitable strategy design should involve the minimum number of steps and methodologies that offer the greatest possible yield. Generally speaking, for analytical scale procedures, high-resolution separation methods can be used for processing small amounts of material. Conversely, for large-scale preparations

Methods for Analysis of Carbohydrate Metabolism in Photosynthetic Organisms.
DOI: http://dx.doi.org/10.1016/B978-0-12-803396-8.00003-X

involving large amount of material, methodologies with good separation capacity but low resolution are usually included in the first stages of purification.

As regards proteins, over the last decade, numerous publications and reviews have addressed the principles of separation and purification methods. The information provided in the first part of this chapter is centered on general considerations and some practical suggestions about the isolation and separation methods most commonly used to study plant, microalgae, and cyanobacterial enzymes. Even though not all proteins are enzymes, to simplify the wording, in this chapter both terms are used interchangeably, considering that the methodologies described can be applied to both of them. The second part of the chapter describes different carbohydrate separation methods that can be applied to studies dealing with photosynthetic organisms.

3.2 Enzymatic Protein Purification

Enzyme purification procedures are addressed to efficiently isolate a given protein with the maximum purity and catalytic activity to conduct further studies (eg, structure analysis, biochemical and biological properties determinations). For this purpose, sufficient amounts of active purified enzymes (several milligrams) have to be obtained. In recent years, molecular biology advances, and the availability of genetic engineering and genomic tools have allowed to produce large amounts of *purified recombinant proteins* that have facilitated the accomplishment of most protein studies. However, enzyme purification from a biological material source remains a key and irreplaceable objective when the authentic protein molecule is needed.

The design of an enzyme purification procedure from a biological material has traditionally been considered as an "art" since it relies on the experience of the researcher and is attained by searching for a scheme of steps by trial and error. Currently, enzyme purification is facilitated by different approaches that take advantage of protein features which led to the development of biospecific methods. Also, the availability of modern instrumentation has helped to achieve protein purification to homogeneity in an easier and faster way. Such is the case of the emergence of fast-flows ion exchange resins and of fast protein liquid chromatography (FPLC). Nevertheless, the "old-fashioned" methods should not be dismissed since they can be successfully used without sophisticated instrumentation and at low costs.

The procedure to be adopted for the purification of a given enzyme implies the selection of biological starting material, the homogenization method (both topics developed in Chapter 2, Preparation of Protein Extracts), and the sequence of the separation methods. Previous data related to the purification of the enzyme of interest (eg, purification from other species, or from tissues of the same species) are generally very useful for the selection.

After cell disruption, the homogenate should be clarified (usually by centrifugation) to obtain the starting solution (crude extract), which, apart from proteins, contains numerous other substances of high and low molecular weight. In analytical scale preparations, the small molecules can be removed by dialysis or gel filtration (eg, using Sephadex G-25 or G-50, or BioGel P10), or it is even possible to remove the small molecules in the first purification step by submitting the protein extract to an ion-exchange or affinity chromatography. However, in large-scale preparations, bulky homogenates are not easy to clarify by centrifugation and preliminary treatments (such as precipitation by salts, organic solvents, pH adjustment, or heat treatment) are generally used to remove particles, unwanted proteins and nonprotein molecules, and to reduce volume. Also, the selective removal of nucleic acids (strongly charged molecules that may not be separated during purification) by addition of streptomycin, dihydrostreptomycin, or protamine sulfate and manganese chloride can be used as a preliminary stage.

The order of the fractionation steps is determined by the starting volume and the principle of the method to be adopted. Importantly, if possible, the initial large volume of the crude extract should be reduced in the first step to maintain enzyme activity, which is usually more labile in diluted solutions. The fractionation procedure should be monitored by measuring enzyme activity after each step, preferentially with a simple and rapid method. To optimize the sequence of methods, each step should be first carried out on a small scale, and then scaled up.

Enzyme separation methods are based on the main protein properties: solubility (eg, pH, solvent, and change in ionic strength treatments, decrease in dielectric constant), size (eg, dialysis, ultrafiltration, fractionation by centrifugation, gel filtration chromatography), charge (eg, fractionation by ion exchange chromatography, electrophoresis, isoelectric focusing, chromatofocusing, hydrophobic chromatography), and structural features (eg, affinity chromatography, affinity elution, dye–ligand chromatography, immunoadsorption, covalent chromatography). The most commonly used methods are described below as a reference guide.

3.2.1 Methods Based on Protein Solubility

As initial stages of a purification procedure, different simple methods are widely used as the preliminary treatment of the extract. Such is the case of protein precipitation by changing the pH, heat treatment, addition of an organic solvent or a salt, which alter the nature of the protein solution.

Fractional precipitation can be carried out by changing the original pH of the protein extract (usually at pH 7.0 ± 1.0). At its isoelectric point (pI), a protein has no net charge, and its solubility is reduced due to the loss of repulsive electrostatic forces between molecules.

Therefore, the adjustment of the protein solution pH can be used either to precipitate a protein of interest or, else, to precipitate unwanted proteins. Importantly, it should be checked if the enzyme under study is not inactivated by the pH modification. Raising the pH above 8.0 is not useful as it does not produce protein precipitation. In many cases, contaminating proteins can be removed by centrifugation after lowering the pH solution to 5.0–5.5 by adding an acid solution (usually with a pH no more than 2 units lower than the final pH) for a few minutes. Acetate buffer (eg, 1.0 M, pH 4.0) is a suitable option to perform this procedure.

The heat treatment is based on the denaturation of less stable proteins at a temperature which does not damage the protein of interest. This is not a generally used technique, particularly for large scale purifications, but it is worth trying with small volumes that can be handled. The procedure must be accurately performed. To illustrate the treatment, the denaturation of a protein could take 10 s at 60°C, while other proteins could be hardly affected after 5 min at this temperature. The increase in temperature of a protein solution should be done in the shortest amount of time possible. The operation can be carried out by heating the solution in a bath at a temperature higher than that selected for precipitation, while stirring; then the protein solution should be kept at the selected precipitation temperature (eg, 60°C) for a short time with continuous stirring and finally rapidly cooled at 0–4°C. The precipitated proteins are removed by centrifugation. Enzyme activity and protein concentration are determined in the supernatant. To adapt this technique to large volumes (liters), efficient mechanical stirring at a controlled speed is required.

The combination of pH adjustment and heat treatment has been successfully applied to better precipitate unwanted proteins and resulted in a fairly good purification level in a short time.

Precipitation of proteins by organic solvents (such as ethanol, methanol, propanol, or acetone) is based on the decrease in the dielectric constant of the solution and the increase in the electrostatic forces. This technique has been used for protein fractionation in large scale purifications as a first-stage treatment, before a chromatographic step. To prevent enzyme denaturation by the addition of the organic solvent, the procedure should be carried out at 0°C or less (eg, −5°C) and the ionic strength of the solution should not be high (electrolyte concentration <30 mM). The organic solvent (kept at −20°C) is added to the aqueous protein solution contained in centrifuge tubes, increasing its concentration in each fractionation step. Depending on the protein and the solvent involved, its concentration could range from 5% to 80% (v/v), at a fixed pH, and ionic strength. This type of fractionation has not been largely accepted, perhaps due to the difficulties entailed in handling solvents. However, its efficiency as an earlier stage of large-scale purification of some proteins has shown to be greater than when ammonium sulfate is used. Polyethylene glycol (M_r 4000–20,000) is an alternative organic compound used as a protein precipitant agent. It is a neutral nondenaturing water soluble polymer that removes water from the hydration spheres of proteins causing precipitation. The increment in polyethylene glycol concentration required to reduce solubility is unique for a given protein–polymer pair. Relatively small amounts of the polymer (such as 5–20% w/v) are needed, which depend on the size of the protein and polymer.

Fractional protein precipitation by salts is a widely applied method, ammonium sulfate being one of the earliest forms of protein purification. Protein solubility is reduced in the presence of increasing salt concentration beyond a certain point. Several factors contribute to this effect, such as the decrease in effective water concentration (by water association in hydrated salt ions), binding of ions to the protein molecule that can generate particular insoluble protein–salt complex, and an increase in hydrophobic interactions between surface protein areas deficient in polar amino acids causing protein aggregation (salting-out). Importantly, the protein suffers a reversible denaturation, since it can be redissolved in buffer after removing the salt. Ammonium sulfate is the most common precipitant used in enzyme purification procedures because of its high solubility in water (c.4 M saturated solution), its low cost, and for not affecting enzyme stability/activity. Salt fractionation is basically an empirical procedure in which proteins are exposed to gradual increments of salt concentration. It is usually carried out as an early step in an enzyme purification scheme. In a preliminary experiment, a suitable ammonium sulfate concentration is added and the precipitate produced is separated by centrifugation. The process is repeated with further salt addition, gradually increasing the concentration in the supernatant solution until most of the enzyme has been precipitated. In each step, enzyme activity is assayed in the pellet and supernatant to verify the fractionation of the protein of interest; and the selection of the salt concentration range is based either on enzyme recovery or on the degree of purification (higher specific activity) results. As an extra benefit, "salting out" in a single step (eg, 0–80%) can be used to concentrate proteins or to clear contaminant material.

3.2.2 Protein Fractionation by Batch Adsorption

These are successful methods based on proteins distribution between a solution and a solid phase. It can be employed either in a batch or in a chromatographic procedure, which offers extra selectivity with a similar adsorbent (see Section 3.2.3.4). The advantage of batch methods is that they are rapid and can deal with large volumes. Briefly, an adsorbent (aluminum oxide such as Cγ alumina gel, titanium oxide, zinc hydroxide gel, calcium phosphate gel, bentonite, ion exchangers, biospecific adsorbents, among others) is added under stirring to the enzyme solution, often the initial extract or the supernatant of an acid or heat treatment. The equilibrium between the adsorbent and the solution occurs in a short time (minutes). Then the absorbent is collected by filtration or centrifugation, and washed. The adsorption process is benefited in slightly acid solutions (pH 5–6) with low salt

concentrations. If the protein of interest becomes adsorbed, it can be separated from other components of the solution after a careful elution protocol using a suitable buffer (often slightly alkaline buffer solutions, as phosphate buffer pH 7.6). On the contrary, if the enzyme is not adsorbed, the method may be useful to remove unwanted proteins.
$C\gamma$ alumina gel and calcium phosphate gel have been the most extensively used adsorbents in fractional adsorption procedures.

3.2.3 Protein Fractionation by Chromatographic Techniques

As mentioned above, the procedures described in Sections 3.2.1 and 3.2.2 are mainly used at the initial stages of purification, and, generally, result in a small increase in the specific activity of the enzyme of interest. For greater purification, and to finally obtain a pure protein, different types of chromatography are usually employed given their greater resolving power and effectiveness for separating proteins.

Essentially, chromatographic techniques are based on the partition or distribution of a compound between two nonmiscible phases: a stationary phase (consisting in an immobilized solid, or gel, or a liquid or solid/liquid mixture) and a mobile phase (a liquid or gas that flows through the stationary phase). In column chromatography, the mode most commonly used for protein separation, the stationary phase is attached to a matrix, packed in a glass, plastic or metal column, and the mobile phase flows through the column, either using hydrostatic pressure or with the aid of a pump system, or applying gas pressure. Matrix material should be inert to minimize nonspecific absorption, insoluble, chemically and mechanically stable, and should permit a good column flow rate. A good matrix could be found in a wide range of particle sizes.

Particularly, column liquid chromatography is one of the most important techniques used for analytical or preparative protein purification, which can be achieved at low pressure as well as by fast protein liquid chromatography. Based on the main protein properties, the separation mechanisms are ion exchange, adsorption, specific affinity to immobilized ligands, and molecular exclusion.

3.2.3.1 Ion-exchange chromatography

Protein purification by ion-exchange chromatography is based on ionic interactions and depends on the electrostatic attraction between species of opposite charge. Thus, buffer-dissolved charged proteins compete for oppositely charged groups on a solid ion exchange adsorbent. It is a very powerful tool, and probably the most extensively used technique for protein separation because it allows an ample range of working pHs, its resolving power is very good and its use is not limited by the sample volume. Importantly, highly reproducible results are obtained when it is scaled-up.

Basically, an ion exchanger is a matrix to which positively or negatively charged groups (anion or cation exchanger, respectively) are covalently bound. Anion exchangers (also known as basic ion-exchangers) attract negatively charged ions; whereas cation exchangers (also known as acidic ion-exchangers) attract positively charged ions.

Matrices commonly used are made of cellulose (eg, anion-exchangers: DE 23, 92, 32, 51, DEAE Sephacel, QA 52; or cation-exchangers: CM 23, 92, 32, SE 92, 52, among others), dextran (eg, anion-exchangers: DEAE Sephadex A-25, A-50, QAE Sephadex A-25, A-50; or cation-exchangers: Sephadex C-25, C-50, among others), agarose (eg, anion-exchangers: DEAE Sepharose CL-6B, DEAE Sepharose Fast Flow, Q Sepharose Fast Flow; or cation-exchangers: CM Sepharose CL-6B, CM–Sepharose Fast Flow, S Sepharose Fast Flow among others), polyacrylamide, polyhydroxyether, copolymers of styrene, and divinylbenzene. These polymers are derivatized with different charged groups that confer the characteristic of the exchanger. On the basis of those functional groups, ion exchangers can be classified as weak or strong.

Both porosity and size of exchanger particles influence the chromatography development. The size of the matrix constituents is defined by the capacity of the particles to pass through a standard sieve and indicated by the mesh number. The higher the mesh number, the higher the surface-volume relationship and capacity of exchanger, and the slower the flow. Matrices with 100–200 mesh and 200–400 mesh are standard materials for analytical scale use, and suitable for high resolution for analytical scale, respectively. Matrices with 50–100 mesh, in turn, are suitable for preparative scale use and exhibit very high flow speed.

The ionic charge of a protein depends on the pH of the solution and its isoelectric point. At a pH below its isoelectric point, the protein is positively charged and will bind to the negative charged group of a cation exchanger. Also, the electrostatic interactions between a protein and the matrix depend directly on the buffer ionic strength. During a purification procedure, the protein in a low ionic strength solution is adsorbed in the selected ion exchanger under conditions that allow a strong binding. Thus, to perform an anion exchange chromatography, the enzyme should be in a buffer solution of about pH 8, containing 10 mM anion; in the case of a cation exchange chromatography, the enzyme solution should be at about pH 5, containing 10 mM cation.

It is worth noticing that charged proteins are reversibly adsorbed to the exchanger and may remain attached to or be eluted from the exchanger by modifying the ionic environment. The more charged the proteins are, the tighter the binding to the exchanger is, and the harder the displacement by another ion becomes. Thus the protein sample is loaded onto the column under conditions in which it will be strongly retained. For elution, bound proteins can be displaced from the exchanger either by modifying the pH, which alters the molecule charge, or by the introduction of counterions into the elution buffer. The increase of the ionic strength of the mobile phase (eluent) is the most frequently used method for eluting proteins, either discontinuously, in a step gradient, or applying a linear gradient that improves the quality of the protein separation. The step gradient is a simple and very effective option for eluting the protein of interest, particularly if prior information on the elution salt concentration is available from a linear gradient experiment. Other types of elution gradients (concave or convex gradient) are more difficult to put into practice.

An ion-exchange chromatography step in an enzyme purification procedure could bring about a 10-fold (or even more) degree of purification, which also explains its wide use.

3.2.3.2 Molecular exclusion chromatography

Molecular exclusion chromatography (also known as gel filtration, size exclusion or molecular sieve chromatography) involves the separation of molecules present in a mixture by their size differences. Gel filtration is usually applied for fractionation of water soluble macromolecules when they pass through a matrix of porous beads made of an insoluble highly-hydrated polymer packed in a column. The packed gel is equilibrated with buffer which fills the pores (liquid referred to as the stationary phase) and the space between beads. The liquid outside the particle is referred to as the mobile phase. Large molecules are unable to enter the pores and are excluded from the liquid within the pores of the beads and emerge first from the column. On the contrary, small molecules can enter the pores and are retarded, passing down the column more slowly than larger molecules. Elution of proteins is performed by keeping the conditions constant throughout the development of the chromatography with a single buffer (isocratic elution mode). The exclusion limit is determined by the porosity of the material. Different pore sizes are obtained by varying the degree of cross-linking of the beaded gel, which determine the range of molecular mass values (M_r) that can be fractionated.

Spherical particles, made of inert, uncharged and rigid hydrophilic polymers of uniform size, constitute the matrix of the material used for gel filtration. The materials most commonly used are Sephadex (cross-linked dextrans whose exclusion limit is determined by the material porosity), BioGel P (cross-linked polymers of acrylamide), Sephacryl HR (cross-linked polymers of dextran and acrylamide, mechanically very rigid), Sepharose (agarose, a natural polymer, made in different particle sizes, thermolabile), Sepharose CL (agarose cross-linked with 2,3-dibromopropanol, rigid and thermostable), and Superose (a highly cross-linked agarose covalently bound to dextran, rigid and thermostable). Sephadex and Biogel are available in several particle sizes (coarse, fine, and superfine) which differ in flow rate and resolution. While superfine materials should be run at lower flow rates and have higher resolution, coarse materials feature higher flow rate and poorer resolution.

Gel filtration is a very useful technique used for protein separation at the end of a purification process, when the partially purified protein is clear and in a small volume.

In this type of chromatography, the sample volume to be loaded onto the column is one of the most important parameters that influence the fractionation resolution. It is expressed as a percentage of the total column volume and should not exceed 70 $mg \cdot mL^{-1}$ protein. Small sample volumes contribute to avoid overlapping of eluted proteins. Manufacturers recommend a sample volume from 0.5% to 4% of the total column volume for high resolution fractionation; however, for most applications it should not exceed 2% to reach maximum separation. Regarding the elution buffer composition, it usually does not affect resolution. The addition of NaCl (at 0.15 M) to the buffer solution is recommended to avoid nonspecific ionic interactions with the matrix.

3.2.3.3 Affinity chromatography

Affinity chromatography is a biospecific technique and one of the most powerful methods for efficient protein purification which takes advantage of a well-known and defined biological property of the molecule to be purified. In the case of enzymes, it is based on a reversible biospecific interaction between the protein of interest and a related molecule or ligand (eg, its substrate or an analog, a competitive inhibitor, a cofactor, a receptor, other specific ligand, or an antibody) covalently linked to an inert matrix, such as agarose. A good matrix for protein affinity chromatography should be chemically stable, inert, rigid, avoid nonspecific interactions, and have large pores. Typical gel supports are 4% and 6% agarose (cross-linked beaded agarose) used for gravity flow only. Also, a copolymer of cross-linked bis-acrylamide and azlactone is available for gravity flow and FPLC systems, at medium pressure.

The ligand/protein interaction can result from electrostatic or hydrophobic interactions, van der Waals' forces and/or hydrogen bonding. In short, a requirement for this type of chromatography is to make available a column with an affinity matrix (ie, gel particles that have the suitable ligand covalently attached to the matrix via a spacer) to which the protein of interest is able to specifically bind. The bound enzyme can subsequently be desorbed from the affinity medium in an active form either specifically, by passing a solution of a competitive ligand over the matrix, or non-specifically, by changing the buffer conditions (pH, ionic strength or polarity) to weaken the interaction between the ligand and the matrix. That is, each specific affinity procedure requires its own setting conditions. Remarkably, this simple, though highly selective method, is capable of purifying an enzyme several thousand-fold in a single step from a crude extract, with high recovery.

The affinity matrix with the suitable ligand may be commercially available or needs to be prepared. Group-specific gels are supplied ready to use and can be applied to purify certain proteins. A wide variety of these products are often supplied by different companies with complete protocols. For example, ligands such as 2′,5′-ADP and 5′-AMP have affinity for enzymes with $NADP^+$ as cofactors, or ligands such as lectins, calmodulin, and benzamidine have specificity for glycoproteins, kinases, and serine proteases, respectively. Particularly,

Concanavalin A (ConA) is a lectin already covalently bound to a matrix such as Sepharose. ConA is able to specifically bind the carbohydrate moiety of some glycoproteins and is commonly used to separate glycoproteins from other proteins. The proteins bound to ConA can be detached by changing the pH, the ionic strength, or by adding sugars that compete for the binding sites on the lectin. On the other hand, the preparation of other covalent-coupling gels requires some ligand-binding chemistry steps before using. Cyanogen bromide (CNBr), tresyl, epoxy, and triazine are the linking groups most commonly used. These gels are available in an activated (ready for binding the ligand) or nonactivated form. The latter require additional activation steps (eg, carbodiimide treatment) prior to ligand binding. Affinity chromatography material can be made from activated agarose precursors (eg, CNBr-activated Sepharose, epoxy-activated Sepharose, and vinyl sulfone agarose) and specific ligands can be bound either directly or via a spacer arm. The gel can be provided with or without a spacer, which allows the availability of the active site of the ligand to the sample. The choice of a spacer depends on the ligand (particularly, for small ligands), the sample, and the linkage chemistry. Experimental instructions for linkage chemistry are usually supplied by the gel manufacturer. The most common issues associated with the implementation of this procedure and with the use of affinity chromatography in enzyme purification have been extensively described in the literature (Ostrove, 1990).

Immobilized metal ion affinity chromatography (IMAC) is another specialized form of affinity chromatography, increasingly used over the last years as a quick, reliable protein fractionation technique. IMAC is based on the specific coordinate covalent bond of amino-acids histidine, cysteine, and tryptophan to transition metal ions (Co^{2+}, Ni^{2+}, Cu^{2+}, Zn^{2+}), through the electron donor groups on the amino-acid side chains. To use this property for protein purification, the metal ion with higher affinity for the matrix than for the protein to be purified must be immobilized onto an insoluble support. Histidine is the amino acid that exhibits the strongest interaction with matrices containing immobilized cobalt or nickel ions. Other column matrices where iron, zinc, or gallium ions are immobilized are used to retain phosphorylated proteins (or peptides).

One of the most extended uses of affinity chromatography is for the purification of recombinant proteins produced as fusion proteins (tagged proteins) to allow purification. Since most proteins do not have affinity for metal ions or other molecules, recombinant DNA technology allows to produce recombinant proteins with a known affinity in order to facilitate their purification. The protein of interest can be produced as a recombinant protein harboring various histidine residues at the amino- or carboxi-terminal ends. This histidine-tagged protein can be retained on immobilized nickel (or cobalt) ions column matrices since the electron donor groups on the imidazole ring of the amino acid readily form coordination bonds with the metal. The protein is eluted by changing the pH or with imidazole (a competitive molecule) either as a gradient of increasing concentration or in a stepwise procedure. The purification of histidine-tagged proteins yields up to 100-fold enrichment in a single step, and 95% purity. Similarly, recombinant proteins produced as fusion with gluthathione-S-transferase (GST) can be purified after retention of the GST region in an immobilized glutathione matrix (commercially available as gluthathione agarose). The tagged protein is eluted with glutathione which displaces it from the matrix.

More recently, the use of immobilized affinity ligands to target biomolecules has extended beyond chromatographic applications. The capture of specific biomolecules by affinity for purification, removal of contaminants, and other analytical uses can be made using ligands coupled to latex beads, nanoparticles, macrobeads, membranes, microplates, array surfaces, dipsticks, and other devices.

3.2.3.4 Adsorption chromatography

Adsorption chromatography, the oldest type of chromatography, is based on protein distribution between a solid material (the adsorbent or the stationary phase) and a solution. The adsorbent has the property of fixing molecules on its surface. In this process van der Waals attraction forces, weak nonionic and hydrogen bonds interactions take place in specific sites, which are able to discriminate between different molecules. During the chromatographic process, those sites are occupied by molecules of the eluent or of the compounds present in the mixture, depending on the relative force of the interactions. Generally, it is more influenced by specific groups rather than by the size of the molecules because the interaction with the adsorption sites is produced by those groups. This technique is mostly used to separate nonionic, water-insoluble molecules. However, crystalline hydroxylapatite [$Ca_{10}(PO_4)_6(OH)_2$] is an adsorbent used to fractionate proteins and nucleic acids. Even though the absorption mechanism remains to be clearly understood, it is believed to involve electrostatic attractions and dipole–dipole interactions between positively charged calcium ions and negatively charged phosphate ions on the hydroxylapatite with protein negatively charged carboxyl groups and positively charged amino groups. The absorption takes place at a neutral pH in 20 mM phosphate buffer, and proteins are eluted by increasing the phosphate buffer concentration up to 500 mM. It is difficult to predict the effectiveness of this type of chromatography based on the properties of the protein of interest.

3.2.3.5 Hydrophobic interaction chromatography

The fact that most proteins have hydrophobic regions or patches on their surface related to the presence of nonpolar amino acids was exploited to develop hydrophobic interaction chromatography (HIC), where the column has a hydrophobic material as the stationary phase. The matrix most used is agarose though other matrices are hydrophilic polymers and polymethacrylate. These matrices can be partially substituted on their surface by different functional groups, such as alkyl (hexyl, propyl, octyl) or aryl groups (phenyl). The alkyl substituents bind more strongly to proteins than the aryl groups do. Examples of HIC column materials purchased by different suppliers are: Phenyl-Sepharose CL-4B, Octyl Sepharose CL-6B, Phenyl Sepharose 6 fast flow, Butyl Sepharose and Octyl Sepharose 4 fast flow (particle size 45–165 μm), Phenyl Sepharose High Performance (with higher resolution than Phenyl Sepharose 6 Fast Flow due to a smaller bead size), Fractogel EMD Pheyl I 650 (S) and Propyl I 650 (S) (particle size 20–40 μm), and Macro-Prep t-butyl and Macro-Prep methyl HIC (particle size about 50 μm).

The binding of a protein to the matrix depends on several factors, such as the degree of matrix substitution, and the nature and size of the hydrophobic regions of the protein. The higher the degree of substitution, the higher the degree of protein binding. The protein solution is loaded onto the column in a solution of high ionic strength (eg, 20 mM potassium phosphate buffer, pH 6.8–7.0, containing 1 M ammonium sulfate). From highest to lowest, the main ions that benefit from binding are: $NH_4{}^+ > Rb^+ > K^+ > Cs^+ > Li^+ > Mg^{2+} > Ca^{2+} > Ba^{2+}$, and $PO_4^{3-} > SO_4^{2-} > CH_3COO^- > Cl^- > Br^- > NO_3{}^- > ClO_4{}^- > I^- > SCN^-$. Therefore, the preferred salts for binding proteins are ammonium sulfate or potassium phosphate. Protein elution from the column is carried out with a decreasing salt gradient (eg, from 1 M to 0 M ammonium sulfate). HIC is often used after ammonium sulfate fractionation, since it is not necessary to remove the salt present in the protein precipitate. Protein elution from the matrix can be facilitated by the addition of glycerol, ethylene glycol or nonionic detergents, such as Triton X-100.

Matrices used in HIC have a very high capacity to bind proteins, a property that can be used for concentrating proteins in diluted solutions.

3.2.3.6 Reverse phase chromatography

In reverse phase chromatography, hydrophobic molecules are adsorbed onto a hydrophobic solid support (ie, the stationary phase is nonpolar) while the mobile phase is polar. The stationary phase is usually silica to which butyl, octyl, and octadecyl groups are chemically bound. The mobile phase is water, buffers, methanol, acetonitrile, tetrahydrofuran, or a mixture of these. The feature of this chromatography is that the stationary phase is essentially inert and only nonpolar interactions (hydrophobic) are possible with the substances to be separated. Consequently, separations are determined mainly by the characteristics of the mobile phase. This mode of chromatography has become increasingly relevant for high resolution separation of proteins and peptides, for instance for purity check analyses. Protein desorption is conducted by adding more organic solvent to the mobile phase decreasing polarity, which reduces the hydrophobic interaction between the solute and the solid support. The higher the degree of protein hydrophobicity, the higher the concentration of organic solvent needed for protein desorption. In view of the fact that organic solvents can denature many proteins, reverse phase chromatography is not recommended for enzyme protein purification.

3.2.3.7 Fast protein liquid chromatography

Fast protein liquid chromatography (FPLC, formerly named "fast performance liquid chromatography") is a form of medium pressure chromatography originally developed for purifying proteins with high resolution and reproducibility. Its distinguishing feature is that the stationary phase is composed of small-diameter beads (generally cross-linked agarose) that are packed in glass or plastic columns and have high loading capacity. Resins for FPLC are available in a wide range of particle sizes and ligand surfaces, which are selected on the basis of their application.

The FPLC system allows the use of a wide range of aqueous buffers (the mobile phase) and different stationary phases to perform the main chromatography modes (ion exchange, gel filtration, affinity, chromatofocusing, hydrophobic interaction, reverse phase). However, anion exchange and gel filtration chromatography are the modes most commonly used.

In general, the mobile phase is an aqueous buffer solution, whose flow rate through the stationary phase is controlled by a pump (normally kept constant), while the composition of the buffer may vary by mixing two or more solutions contained in external reservoirs. In most common FPLC strategies, eg, ion exchange, the resin is selected in a way that the given protein can be bound to the resin through charge interaction in a buffer A (running buffer), and subsequently, it can be dissociated and taken back to the solution in a buffer B (elution buffer). In contrast to high-performance liquid chromatography (HPLC), the buffer pressure used is low, typically 5 bar, but the flow rate is high (eg, $1-5\ \mathrm{mL \cdot min^{-1}}$). FPLC chromatography can be scaled up, allowing the analysis of samples containing from milligrams of proteins in 5 mL-columns to preparative production of kilograms of purified proteins using columns of several liters of volume.

3.2.3.8 Isoelectrofocusing

Isoelectrofocusing (IEF) is an electrophoretic technique that enables the separation of proteins on the basis of their isoelectric points. Proteins (as well as amino acids and peptides) are amphoteric molecules that contain both positive and negative groups and their net charges are determined by the pH of their surroundings. Thus, proteins exhibit their net charge based on the type and number of ionizable groups located at the side chains of their amino acids and prosthetic groups.

Protein separation by IEF relies on the movement of molecules due to a potential difference in a pH gradient. At pH values below their isoelectric point, proteins are positively charged and move toward the cathode during electrophoresis. On the contrary, at pHs above their isoelectric points, proteins are negatively charged and migrate toward the anode. The region corresponding to the anode is acidic while that of the cathode is alkaline.

Migration leads proteins to a region where the pH matches their isoelectric point. At this time, the protein does not move in the electric field because its net charge is zero. As a result, proteins will focus, at different rates, on sharp stationary bands, in a position where the gradient pH is coincidental with their isoelectric point. Proteins can remain at those pH values for extended periods. When the electric equilibrium point has been reached, the proteins present in the sample are separated based on their isoelectric points. Then they can be fractionated by a conventional method and detected, eg, by measuring enzyme activity.

3.3 Carbohydrates Separation

The methodologies for carbohydrate separation and identification have been essential to study cell metabolism and necessary for the analysis of natural sugars and the identification of substrates (or products) of enzymatic reactions. Different modes of classical chromatography have been adopted, starting with paper chromatography, following with thin layer chromatography (TLC), gas–liquid chromatography, and HPLC. Also new methodologies that require more sophisticated equipment (such as mass spectrometry (MS)) are now being used for the analysis of carbohydrates. The most useful and classical types of chromatography are described below.

3.3.1 Partition Chromatography

Partition chromatography is based on the nature of the components present in the sample to be analyzed. The molecules get preferential separation between two phases (stationary phase and mobile phase) which are in contact (see Section 3.2.3). The components of the mixture are distributed into both phases during the flow of the mobile phase based on their partition coefficients in each phase. The partition coefficient of a substance is defined as the relationship of its concentration in each of the phases. Partition chromatography can be used to separate molecules of low molecular weight, such as monosaccharides, disaccharides, oligosaccharides, and sugar nucleotides, either through columns using hydrostatic pressure or with the aid of a pump system, or applying gas pressure. Partition is the principle of separation of different chromatographic types, including paper chromatography, HPLC, gas chromatography, and high performance thin layer chromatography (HPTLC). Other techniques are based on the principle of adsorption.

3.3.2 Paper Chromatography

Chromatography on sheets or strips of filter paper constitutes an important separation method and standard practice for the investigation of sugars and other metabolites. Despite the fact that this method does not yield accurate quantitative results and has been largely replaced by TLC, this powerful tool has the special advantage of exploiting very small differences in partition coefficient which leads to good separations. In addition, it is inexpensive and calls for small amounts of compounds. Among the different types, descending paper chromatography has been the most extensively used.

In some cases, the sample should be previously desalted because salt affects the separation quality. Desalting can be achieved by passing through columns of Dowex-50 and Amberlite IR4B-OH. It should be underlined that strong anionic columns cannot be used because they could absorb sugars. This step implies an increase in volume, so further freeze-drying is needed.

Briefly, the method consists of loading a small quantity of the sample to be analyzed in a circular spot near the top of a sheet or strip of filter paper which is hung vertically from a trough containing the developing solvent. The top edge of the paper is immersed in the solvent, and placed inside a chromatographic tank saturated with the solvent. The liquid penetrates the paper by capillarity, passes over the spot, and advances down for several hours. Finally the paper is dried and the position of the separated components is determined by different developing reagents.

Different sugars can be separated using Whatman chromatography papers N°1 or N°4 for analytical separation, or Whatman paper N°3 for preparative separations. The difference between Whatman N°1 and N°4 is in the speed of development, N°4 being faster than N°1, but separations are not as good as with N°1. The solvents most commonly used for the separation of mono- and oligosaccharides are butanol:pyridine:water (6:4:3, v/v/v) and ethylacetate:pyridine:water (12:5:4, or 8:2:1, v/v/v), and to separate mono- and disaccharides, water-saturated phenol (phenol/water (4:1, v/v). The sugar positions on the paper are ascertained by comparing with the position of standards after using a suitable developing reagent (eg, silver nitrate or naphthoresorcinol reagent).

3.3.3 Thin Layer Chromatography

Carbohydrate analytical separation can also be obtained by TLC. In this technique, cellulose or silica plates replace paper and the amount of sample that can be analyzed is about 10 times smaller than that used in paper chromatography. Similar precautions to those for paper chromatography with regard to the sample salt content must be taken. Also, the mixture of propanol:ethylacetate:water (6:1:3, v/v/v) can be used for the separation of mono- and oligosaccharides, and n-butanol:isopropanol:water for the separation of oligosaccharides.

3.3.4 Carbohydrate Detection on Chromatograms

The position of different carbohydrates either on paper or thin-layer chromatograms can be visualized with various reagents. Two of the most widely used are described below.

3.3.4.1 Silver reagent

The silver reagent consists in a silver nitrate saturated solution and an alcoholic solution of NaOH. The procedure for preparing these solutions is as follows:

Solution A: Add 0.1 mL of $AgNO_3$ saturated solution (3 g of $AgNO_3$ in 1 mL of distilled water) to 20 mL of acetone. To this solution, add distilled water, drop by drop, with gentle stirring until the precipitate is completely dissolved.
Solution B: Mix 1 mL of 10 N NaOH and 20 mL of ethanol.

Carbohydrate detection is carried out by passing the paper through Solution A or spraying Solution A onto the thin layer chromatogram. Allow paper or thin layer chromatogram to dry at room temperature until acetone is no longer perceived. Once the support is dry, pass it through Solution B. Carbohydrates appear as black spots on a brown background. Monosaccharide spots appear immediately, but nonreducing disaccharides require that the paper or thin-layer chromatogram be exposed to water steam.

3.3.4.2 Benzidine reagent

Benzidine reagent has been used for the detection of reducing compounds. The procedure for preparing the reagent is as follows:

Benzidine stock solution: Dissolve 1 g of benzidine in 40 mL of glacial acetic acid, heating slightly if necessary. Dissolve trichloroacetic acid in 40 mL of distilled water. Mix both solutions. The resulting mixture is stable at 4°C. Keep the mixture in a caramel color bottle. For working solution, dilute 1 volume of stock solution in 9 volumes of acetone.

To detect carbohydrates, pass the paper through the benzidine reagent or spray benzidine reagent onto the thin layer chromatogram, and allow drying at room temperature. Subsequently, heat the paper at 100–110°C until the appearance of stained spots. Each carbohydrate develops a different color.

3.3.4.3 Naphthoresorcinol reagent

Naphthoresorcinol-HCl reagent has been used for the detection of ketoses. The procedure for preparing the reagent is as follows:

Naphthoresorcinol (100 mg) is dissolved in a mixture of 2 N HCl (20 mL) and ethanol (80 mL).

To detect fructose-containing carbohydrates on a paper chromatogram, spray the paper and heat at 75–100°C. Fructose shows with a bright red color, aldohexoses do not react (colorless).

3.3.5 High-Performance Liquid Chromatography

HPLC is a powerful analytical tool that has a number of advantages, such as high resolution and sensitivity in short time, generation of quantitative results, a high potential for automation, a wide range of applications, and good reproducibility. On the other hand, this technique also has some drawbacks mainly related to the high cost of the instrumentation and operation.

The key difference between HPLC and chromatography developed at normal pressure is that the stationary phase particles are very small (5–10 μm) and are packed in stainless steel columns. Differences in resolution of the components of a mixture can be obtained using two different sizes of matrix particles, or with different flows.

The smallest particle size of the stationary phase results in a high resolution of mixtures to be separated, and requires that the system works at high pressure. This facilitates the movement of the mobile phase, and also accelerates chromatography times (eg, 2 $mL \cdot min^{-1}$ for a column 4 mm in diameter and 250 mm long). In other words, HPLC provides double benefits: high resolution and shorter chromatographic times.

HPLC does not differ significantly from low-pressure chromatography, but some special requirements should be met. Usually, already packaged columns are used, and their sizes vary for analytical use (eg, 4 mm × 125 mm or 4 mm × 250 mm with internal diameter bore of 4.5 mm, or small microbore columns with internal diameters of 1.2 m) or for developing preparative chromatography (usually 25 mm × 250 mm or 50 mm × 250 mm). These columns are expensive and extra care is essential. In addition, particular care should be taken with the preparation of the sample and solvents since trace impurities or presence of air could interfere with the detection systems and generate false peaks during elution.

The selection of the mobile phase depends on the type of separation to be achieved. Isocratic separations can be developed with one pump, using a simple solvent, or two or more premixed solvents in fixed proportions. On the other hand, elution through gradients often requires the use of independent pumps in order to deliver both solvents in predetermined proportions through a programmable or preset gradient. The pumps that generate pressure to the mobile phase are the key feature of the HPLC system. The output capacity of a good pump system should be between 6000 and 9000 psi and, ideally, a continuous flow (generally flow capacity of less than 10 $mL \cdot min^{-1}$ and above 100 $mL \cdot min^{-1}$ for preparative separations). The sample is loaded onto the column by the injection of the sample with a micro syringe or by using an injector loop, a key procedure to obtain a successful separation during chromatography.

It is difficult to carry out carbohydrate separation by common chromatography since these compounds have similar physical and chemical properties and there is no suitable chromophore for their detection. Other methods have been developed using the HPLC technique with strategies based on different detectors. Pulse amperometric, electrochemical, and fluorescence (only when the sample is derivatized with a suitable fluorescent tag) detectors as well as detection in charged aerosol, and MS have been used.

The first HPLC columns for carbohydrates allowed mono- and disaccharide separations by eluting with water and acetonitrile:water, and following the elution by refractive index. By contrast, the separation of mono- and oligosaccharides, sugar acids, such as sialic acids, sugar alcohols, sugar phosphates, and sugar nucleotides, based on their ionization can be attained by using high-performance anion-exchange columns and isocratic elution with a highly alkaline solution (pH 12–14) as a mobile phase. The elution is followed by a pulsed amperometric detector (PAD).

Carbohydrate separation by HPLC has been successfully coupled to MS. The combination of these two techniques is a powerful tool used for many applications directed to the detection and identification of compounds (analytes) in a complex mixture. The main benefits are very high sensitivity, excellent selectivity, and the simultaneous acquisition of information about the mass of each analyte, reducing misidentification.

Further Reading and References

AKTA FPLC, System Manual, Amersham Pharmacia Biotech. Edition AB, vol. 18, pp. 1140–1145.
Ardrey, R.E., Ardrey, R., 2003. Liquid Chromatography-Mass Spectrometry: An Introduction. J. Wiley, London.
Block, R.J., Durrum, E.L., Zweig, G., 1955. A Manual of Paper Chromatography and Paper Electrophoresis. Academic Press, New York, pp. 484.
Bornhorst, J.A., Falke, J.J., 2000. Purification of proteins using polyhistidine affinity tags. Methods Enzymol. 326, 245–254.
Chicz, R.M., Regnier, F.E., 1990. High-performance liquid chromatography: effective protein purification by various chromatographic modes. In: Deutscher, M.P. (Ed.), Guide to Protein Purification, Methods Enzymology, vol. 182. Academic Press, San Diego, CA, pp. 392–421.
Cutler, P., 1996. Size-exclusion chromatography. Methods Mol. Biol. 59, 269–275.
Dixon, M., Webb, E.C., 1979. In: Dixon, M., Webb, E.C. (Eds.), Enzymes. Longman Group Limited, London.
Dong, M.W., 2006. Modern HPLC for Practicing Scientists. John Wiley & Sons, Inc.
Freifelder, D., 1982. Physical Biochemistry. Applications to Biochemistry and Molecular Biology. W.H. Freeman and Company, New York.
Holme, D.J., Hazel, P., 1993. Analytical Biochemistry. Longman Scientific & Technical, New York.
Ingham, K.C., 1990. Precipitation of proteins with polyethylene glycol. Methods Enzymol. 182, 301–306.
Kobata, A., Endo, T., 1992. Immobilized lectin columns: useful tools for the fractionation and structural analysis of oligosaccharides. IEEE Audio Electroacoust. Newslett. 597, 111–122.
Kennedy, R.M., 1990. Hydrophobic chromatography. In: Deutscher, M.P. (Ed.), Methods Enzymology, vol. 182. Academic Press, San Diego, CA, pp. 339–343.
McMaster, M.C., 2005. LC/MS: a practical user's guide. John Wiley, New York.
Ostrove, S., 1990. Affinity chromatography: general methods. In: Deutscher, M.P. (Ed.), Methods Enzymology, vol. 182. Academic Press, San Diego, CA, pp. 357–371.

Pingoud, A., Urbanke, C., Hoggett, J., Jeltsch, A., 2005. A Concise Guide for Students and Researchers. Biochemical Methods. Wiley-VCH Verlag GmbH, Weinheim.
Pontis, H.G., Blumson, N.L., 1958. A method for the separation of nucleotides by concave gradient elution. Biochim. Biophys. Acta. 27, 618–624.
Rossomando, E.F., 1990. Ion-exchange chromatography. In: Deutscher, M.P. (Ed.), Methods Enzymology, vol. 182. Academic Press, San Diego, CA, pp. 309–317.
Sheehan, D., 1996. Fast protein liquid chromatography. Methods Mol. Biol. 59, 269–275.
Sheehan, D., 2002. Physical Biochemistry: Principles and Applications. John Wiley and Sons, LTD, Baffins Lane, Chichester, West Sussex, UK.
Sheehan, D., Fitgerald, R., 1996. Ion-exchange chromatography. Methods Mol. Biol. 59, 145–155.
Sheehan, D., O'Sullivan, S., 2003. Fast protein liquid chromatography. Protein Purification Protocol., 244–253.
Snyder, L.R., Kirkland, J.J., Dolan, J.W., 2010. Introduction to Modern Liquid Chromatography, third ed. John Wiley & Sons, Inc.
Stellwagen, E., 1990. Gel filtration. In: Deutscher, M.P. (Ed.), Methods Enzymology, vol. 182. Elsevier Inc., pp. 317–328.
Tswett, M., 1896. Bulletin de Laboratoire de Botanique Générale de l'Université de Genève, vol. 1.
Westermeier, R., 2001. Isoelectric focusing. Methods Mol. Biol. 59, 239–248.
Wilson, K., Walker, J., 1994. Principles and Techniques of Practical Biochemistry. Cambridge University Press.

CHAPTER 4

Measurement of Enzyme Activity

Chapter Outline

4.1 Introduction

It is well established that cell metabolism is governed by enzymes, macromolecules that act as chemical catalysts enhancing the rate of a chemical transformation. With a few exceptions (ie, catalytic RNA molecules), enzymes are highly specialized proteins whose characteristic property is their capacity to catalyze definite reactions essential to living systems. The amount of an enzyme in a cell (or tissue) extract or in a purified (or partially purified) enzyme solution can be measured in terms of the catalytic effect that the enzyme produces (ie, by determining the increase in the rate at which a substrate is converted to a product).

In the field of carbohydrate metabolism of plants, unicellular algae, and cyanobacteria many aspects involving enzymes require further investigation, such as the characteristics and properties of enzymes, the regulation of metabolisms, metabolic flux analyses, or even, the discovery of new carbohydrates and pathways. It is worthy of mention that many enzymes involved are mostly exclusive of those organisms and only a small part of their characteristics (mainly biochemical, biological, thermodynamical, kinetic, and structural properties) has been investigated.

This chapter describes basic principles and some general particularities for the assay of enzyme activities in crude protein extracts, as well as in partially purified or pure enzyme preparations, useful tools in the study of carbohydrate metabolism in oxygenic photosynthetic organisms.

Methods for Analysis of Carbohydrate Metabolism in Photosynthetic Organisms.
DOI: http://dx.doi.org/10.1016/B978-0-12-803396-8.00004-1

4.2 Measurement of Velocity

The objectives of measuring enzyme activity are numerous, including among others, the investigation of biochemical properties of a given enzyme, its biological and physiological significance in an organism (such as to elucidate its role in the metabolism and in the adaptation to the environment), its occurrence among species and/or tissues and its origin and evolution.

Enzyme activity is a measure of the quantity of active enzyme present in the preparation to be tested. Regardless of the purpose of the study, in all cases, enzymatic activity data need to be comparable between samples and between investigators. As results are dependent on temperature, pH, substrate concentration, presence of ions, etc., these assay conditions should be specified. To prevent erroneous results and to obtain accurate and reproducible ones, it is important to revise the concept of reaction velocity, a measure of the conversion of substrate to product that indicates how fast the reaction takes place in a unit of time under in vitro defined conditions (see Section 4.3).

An enzyme reaction progress curve can be graphed by the appearance of a product (or the consumption of a substrate). As depicted in Fig. 4.1, at initiation of the reaction, the product generation (or substrate loss) per time unit is rapid and the variation of substrate transformed is proportional to the time interval. However, as the reaction proceeds, the velocity approaches zero, which can result from one or more reasons, such as product

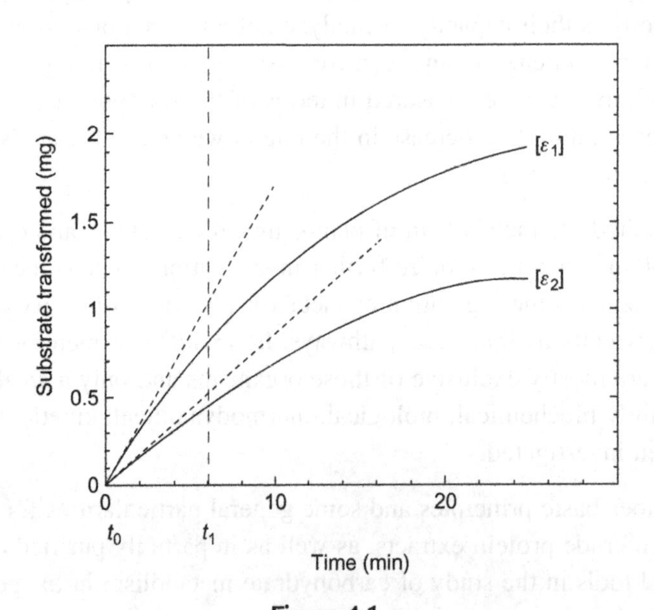

Figure 4.1
Progress curve of an enzymatic reaction showing the appearance of the product, at two enzyme concentrations (ε_1 and ε_2). $[\varepsilon_2] = 2 \times [\varepsilon_1]$.

inhibition, a decrease in enzyme saturation by substrate (due to substrate conversion), an increase in the reverse reaction (as product concentration increases), and/or enzyme (or a coenzyme) inactivation. Consequently, the curves do not often fit the standard equations of homogeneous chemical reactions. Therefore, when studying an enzyme reaction, initial velocities should be measured, which correspond to the initial part of the progression curve (in practice, they are obtained from the slope of the linear part of the plot). In enzyme activity determinations, it is essential to ensure measurements of initial velocities, which commonly correspond to $< 5\%$ of product conversion. However, for some enzymes, the time course of a product appearance fits a linear function up to the time when about 10–20% of the initial substrate concentration has been converted to product. Also, initial velocities should be proportional to the amount of enzyme to avoid misleading results.

The rate of reaction produced by a given amount of an enzyme preparation can be expressed as the amount of product formed (or substrate converted) per time unit. Enzyme activity is often expressed as $\mu mol.min^{-1}$, and one enzyme unit is defined as the amount of enzyme that catalyzes the transformation of a mass unit of substrate per time unit under defined conditions of pH and temperature. The amount of activity per unit of enzyme (units.mg $protein^{-1}$) is called the specific activity. In accordance with the International System of Units, the enzyme unit adopted is the katal (kat), where 1 kat is the amount of enzyme that converts 1 mole of substrate per second. Therefore, 1 $\mu mol.min^{-1}$ is equivalent to 16.67 nkat.

4.3 Enzyme Assay Conditions

While the maximum rate of an enzymic reaction is an intrinsic property of the protein molecule, the in vitro measurements depend on the reaction conditions that are usually different from those present in vivo. Therefore, enzyme assay conditions should be well-defined so that the results are consistent and reproducible.

Enzyme activity is affected by the reaction mixture components (mainly by substrate concentration, pH, ionic strength, and salts) and the temperature. Routinely, enzyme activity is measured at a fixed temperature (eg, 20–30°C are often used for plant enzymes). With respect to pH, ideally, activity should be assayed at the enzyme optimum pH, though, in practice, the buffer added to the reaction mixture may have a pK value ± 1 of the reaction pH, and a concentration greater than that of any other ionizable compound present. It should be considered that some buffers are very sensitive to temperature (eg, Tris and other nitrogen containing buffers) or to ionic strength (eg, phosphate-containing buffers, which change their pH after dilution). Additionally, salts, osmolytes, and reducing agents (eg, NaCl or KCl, polyethylene glycol, glycerol, or sucrose, β-mercaptoethanol, or dithiothreitol, respectively) are usually added to incubation mixtures to maintain enzyme stability and activity.

In routine enzyme activity assays, it is particularly relevant that the substrate concentration at the reaction mixture leads to reproducible and comparable results. The K_m for the substrate (or for each substrate, in the case of a two- or more substrate reaction), the solubility, and the possibility that high substrate concentrations may inhibit the reaction should be contemplated. To assure V_{max} conditions in the case of Michaelian enzymes, the substrate concentration used in a routine assay should be several times the K_m value ($\geq 2 \times K_m$). The closer to K_m the substrate concentration is, the interfering reactions achieved by other enzymes present in nonpurified protein preparations may reduce substrate levels and slow the rate of the reaction. Regarding allosteric enzymes, general conditions of assay cannot be provided, and so they should be analyzed on a case-by-case basis.

For ready reversible reactions, the enzyme activity can be measured in a direction opposite to that occurring in vivo. For example, the activity assay for sucrose synthase, which catalyzes a reversible reaction (Sucrose + UDP (or ADP) $\leftrightarrow$ Fructose + UDP-glucose (or ADP-glucose)), is frequently carried out in the sucrose synthesis direction, which has been shown not to be the reaction direction in vivo in most organisms.

4.4 Enzyme Assay Methods

The assay of an enzyme activity requires an appropriate method to measure how much substrate has been consumed, or how much product is formed during a time interval in the presence of the other in the reaction mixture. Generally, product determination is preferable (even if it is in a small amount) because its concentration increases from zero and its appearance is a net measure of the reaction progress. In contrast, measuring the small decrease in substrate concentration introduces errors and requires a very sensitive and reproducible method for quantitation.

Assay methods have been traditionally classified into three types: continuous, coupled, or discontinuous. They differ with regard to the matter of separation of the product from the substrate, or the requirement of a termination step. In a continuous assay the progress of the reaction is followed as it occurs, hence a separation step prior to detection is not necessary. Such assays are usually preferred because the enzyme action produces a change in a readily detectable physical parameter (eg, fluorescence, absorbance, viscosity, pH) and the result is immediately obtained. A typical continuous assay is based on the registration of changes in absorbance, as it happens when NADH/NAD^+ (or NADPH/$NADP^+$) interconversion takes place and can be followed directly in a spectrophotometer cuvette, by measuring the change in optical density at 340 nm.

In the case of certain enzymes whose substrates and products do not produce any detectable parameter change for measurement of their concentration (eg, spectrophotometrically or spectrofluorometrically), the product generation can be paired with a second enzymatic reaction and the ultimate product formed may be monitored by a continuous assay. Such coupled assays offer the advantage that the product is removed avoiding inhibition by product accumulation and the reverse reaction proceeds. As a way of example, the assay of the enzyme hexokinase that catalyzes the following reaction: Glucose + ATP → Glucose-6-phosphate + ADP can be mentioned. The product glucose-6-phosphate can be oxidized to 6-phospho-gluconate in the presence of oxidized nicotinamide adenine dinucleotide (NAD^+) by the coupled reaction catalyzed by glucose-6-phosphate dehydrogenase. During this oxidation, an equimolar amount of NAD^+ is reduced to NADH. The consequent increase in absorbance at 340 nm is directly proportional to glucose concentration.

A third group of methods for measuring enzyme activity that require two steps are referred to as discontinuous assays and involve stopping the reaction after a fixed time followed by the quantitation of product formed by any method (either chemical or enzymatic).

Often a selective method can distinguish between substrate and product with no separation step, and any of the methods described in Chapter 1, Determination of Carbohydrates Metabolism Molecules, can be used for the determination of products of carbohydrate metabolism enzymes (such as monosaccharides, disaccharides, inorganic phosphate, sugar nucleotides, etc.). To stop the reaction, any denaturing agent (heat, a reagent corresponding to a quantitation method, addition of an acid or an alkali, a detergent, heavy metal ions as inhibitors, etc.) or a dilution with a suitable buffer can be applied. In other enzyme reactions, separation of the remaining substrate(s) from product(s) is required (eg, when radioactive substrates are used). Charcoal precipitation, thin layer chromatography, passage through ionic resins, or high-performance liquid chromatography, are among the general separation methods employed.

Diverse detection methods have been used to quantitatively follow enzyme reactions. Spectrophotometric and spectrofluorometric methods are the most commonly used due to their versatility, simplicity, and rapidity. Also radioisotopic measurement can be employed using radiolabeled substrates in the incubation mixture. This strategy requires an appropriate technique for separating the substrate from the labeled product. Even though the radioactive methods are entering disuse for routine enzyme measurements, as they require special safety conditions and equipment, in some cases, they are irreplaceable given their high sensitivity.

4.5 General Considerations for Planning an Enzyme Assay

The design of an enzyme assay depends on the previous knowledge of the enzyme reaction and properties, and the metabolism in which it is involved. If that information arises from the enzyme originated in the organism under study, the next step is to carry out the preparation of the reaction mixture, and the detection and quantification of the activity as it has been previously described.

Conversely, if no previous reports are available from the organism of interest, and the enzyme activity is assayed for the first time in a protein extract (or in a partially purified preparation) from a given cell, tissue, or organ, the assay conditions should be adjusted after preliminary trial and error experiments. Later basic biochemical studies of the enzyme are necessary (ie, determination of substrate specificity, optimum pH, kinetic parameters, molecular properties, etc.) to design an appropriate assay method for routine enzyme activity determinations.

Importantly, different sources of biological material imply dealing with interfering secondary reactions that can act on the same substrate or on the product of the reaction under study. These reactions can be evidenced by designing appropriate and wise controls, and should be eliminated or minimized to obtain reliable activity data.

For example, in the measurement of sucrose-phosphate synthase (SPS) activity in plant leaf or seed extracts by the quantification of UDP or sucrose (by the thiobarbituric acid method), the presence of phosphatases and sucrose synthase (SuS) causes an overestimation of the activity since fructose-6-phosphate (substrate of SPS) is dephosphorylated to fructose, which is a substrate of SuS. The secondary reactions can be minimized by adding a phosphatase inhibitor (such as NaF) and arbutin or phenyl-S-glucoside, two phenolic glycosides that are inhibitory to sucrose synthase activity for sucrose formation.

After the components of the reaction mixture are defined, and usual controls are designed (ie, enzyme and substrate are separately omitted in the reaction mixture), an analysis method and a detection system should be selected. Based on the type of method, it may be necessary to include other reaction controls. Prior to the initiation of the reaction, the amount of enzyme, the time interval of sampling (for discontinuous methods), and the volume of reaction should be determined. These decisions are taken after a trial and error process. Finally, to carry out the activity assay, the order in which the reaction mixture components are added should be especially considered. In general, the reaction is started by the addition of the enzyme preparation to mixture reaction.

Further Reading and References

Copeland, R., 2000. Experimental measures of enzyme activity, Enzymes: A Practical Introduction to Structure, Mechanism, and Data Analysis. second ed. Wiley-VCH, Inc., New York (Chapter 7).

Dixon, M., Webb, E.C., 1979. Enzyme techniques, Enzymes. third ed. Longman, London (Chapter 2).

Price, N.C., Stevens, L., 1989. Fundamentals of Enzymology. third ed. Oxford University Press, Oxford.

Harris, T.K., Keshwani, M.M., 2009. Measurement of enzyme activity. In: Burgess, R., Deutsche, M. (Eds.), Guide to Protein Purification, Methods Enzymology, vol. 463. Academic Press, San Diego, CA, pp. 57–71.

CHAPTER 5

General Introduction to Mass Spectrometry and Nuclear Magnetic Resonance

Chapter Outline

5.1 Introduction

Mass Spectrometry (MS) and Nuclear Magnetic Resonance (NMR) spectroscopy are among the modern analytical chemistry techniques applied in the study of biochemical and molecular biological problems (Macomber, 1998; De Hoffmann and Sroobant, 2007). In carbohydrate studies, MS helps to identify and quantify different compounds and NMR has become the preeminent technique for structural and conformational analysis of saccharides, glycoconjugates, and other derivatives. These techniques are usually implemented in specialized laboratories, equipped with specific sophisticated and expensive instruments. The fundamental bases underlying both methods and a few applications are described below.

5.2 Mass Spectrometry

MS is a high-power analytical technique used to identify and quantify a chemical system based on the molecular weight of its constituents. It can be applied to a broad range of known or unknown substances, either as pure compounds or mixtures, of an inorganic, organic, or biological nature, in a solid, liquid, or gas state (Herbert and Johnstone, 2003; Jürgen, 2004). This technique also has the advantage of having such a high resolution level that enables the carrying out of specific studies on stereochemistry as well as

Methods for Analysis of Carbohydrate Metabolism in Photosynthetic Organisms.
DOI: http://dx.doi.org/10.1016/B978-0-12-803396-8.00005-3

analyses of a wide range of molecular species based on their chemical properties. It is possible to detect compounds at very low concentrations (pmoles) in complex mixtures and likewise to obtain data from molecular weights to their three-dimensional structures (Watson and Sparkman, 2007).

While there are a variety of methodologies available, in practice the options are reduced to a few which are best suited to answer the usual questions in an analytical chemistry laboratory: "what?" and "how much?" Finally, the choice depends mainly on the stability of the sample.

MS is not considered a typical spectroscopic method since there is no need of any type of radiation to generate spectra. Instead, this technique is based on the ionization of the species under study. MS consists of four instrumental stages: introduction of the sample, ionization, analysis of generated ions, and detection.

The first stage involves the entry of the vaporized sample to the ionization chamber of the spectrometer (ionization of the sample molecules). This is easily achieved in the case of organic substances due to the vapor pressure that they develop when heated. Gas chromatography equipment is an excellent generator of pure substances in the gas phase, because they separate volatile components from complex chemical mixtures. The sample can also enter the vacuum chamber by direct introduction within a small container.

In the next stage, the gas molecules are ionized in the vacuum chamber by the collision of an electron beam. Samples that cannot be either heated or vaporized (eg, proteins and carbohydrates) can be vaporized and ionized in a single step in a technique called electrospray ionization (ESI). In this process, the sample solution passes through a metal nozzle connected to a high voltage generator. The combination of the liquid flow exiting the nozzle with the occurrence of electrostatic charges on the spray droplets (due to nozzle voltage) generates a cloud of positively charged macromolecules that are directed to the analysis stage (see below). A soft ionization technique like MALDI (Matrix-Assisted Laser Desorption-Ionization Molecules) can be used with molecules (such as sugars, DNA, proteins, or other biological macromolecules) which have the tendency to be fragile when ionized by a conventional method. In the MALDI technique, the sample is embedded into a matrix of a solid organic compound that absorbs ultraviolet (eg, by co-dissolution of both substances in the same solvent and subsequent evaporation of the solvent). A UV laser pulse of short duration heats the organic matrix. The vaporization of the matrix carries the embedded molecules towards the vapor phase. It was shown that at high temperature, vaporization is much faster than decomposition of the sample (ie, the laser heating vaporizes the mixture without decomposing the sample).

The molecules ionized by any of the three methodologies described above are electrostatically forwarded to the next stage. With small instrumental modifications, some of the ions generated are able to acquire negative charge, which allows a very sensitive quantification. Usually, the negative-ion spectrometry features much lower signals than the positive-ion spectrometry.

The ion analysis is performed by various techniques, electrostatic, time-of-flight, and Fourier-transform being the most commonly used. In the electrostatic analysis, a device called quadrupole retains all ions generated in the previous step except those with a specific mass range, determined by the device operating parameters. In practice, this device routinely ensures a selection of an ion mass range within ± 1 mass unit (the heaviest and lightest ions are discarded inside the quadrupole). In time-of-flight analyzers, the sudden application of a voltage inside the device drives the charged molecules to travel across a long vacuum chamber, like in a race. Assuming (in the best case) that electrostatic charge is the same for all ions, the kinetic energy transferred after the voltage pulse is the same in each particle. Therefore, the different ion masses translate into different traveling speeds and, thus, into different travel times (time-of-flight). This technique allows a mass accuracy of 10^{-5} units. In Fourier-transform techniques, the electrostatically-accelerated ion jet is subjected to a strong magnetic field, which causes the ions to rotate in circular paths according to the mass/charge ratio. The radiofrequency signals generated by these movements are analyzed by the Fourier transform to obtain a mass spectrum with an accuracy of 10^{-6} units.

The detection stage involves a multiplier tube similar to the photo tubes used in UV-visible spectroscopy. In this case, the impact of ions on the receiving tube triggers a stream of electrons proportional to the number of ion impacts and generates the signal to be analyzed. This detection methodology is extremely sensitive and, therefore, MS is suitable for the analysis of trace components. However, the whole process destroys the sample.

The three vaporization/ionization methodologies can be combined with the three described ion analysis stages to suit the specific type of required analysis, although some combinations of them are instrumentally more usually applied than others. For example, electron ionization followed by a quadrupole is preferred for small organic molecules, which are vaporized and separated by gas chromatography. For larger molecules, usually separable by liquid chromatography, the selected methodologies usually include ESI combined with a time-of-flight analysis.

In view of the fact that carbohydrates and proteins are heat sensitive compounds, a previous derivatization is required to make them more volatile. This allows their separation in a gas chromatograph (short fragments with total mass less than 1000), or after an ESI followed by a time-of-flight analyzer (for masses greater than 1000). Remarkably, the ion analysis techniques with higher resolution can distinguish between two molecules classified as mass isomers.

5.3 Applications of Mass Spectrometry

MS has become an extensively used analytical technique for analyses of compound mixtures or purified biomolecules, including carbohydrates. Modern mass spectrometers make possible the analysis of compounds in the femtomole range and of high molecular mass molecules (over 100 kDa). It can be applied to molecular mass determination and structure elucidation. In the carbohydrate field, a multidimensional approach is required to achieve a comprehensive analysis of mixtures from biological matrices that display a high degree of compositional and structural heterogeneity (Rodrigues et al., 2007). ESI is the technique most frequently used in MS to produce ions since it is an extremely soft method, suitable to ionize highly polar compounds, and imparts relatively low energy to the samples, producing less fragmentation during ionization. By ESI, ions can be formed with positive or negative characteristics depending on the sample. For example, oligosaccharides containing acidic groups (such as sulfate, carboxylate or phosphate) are easily analyzed using the negative-ion mode.

MS can also be applied to the study of enzymatic reactions (eg, protein structures, enzyme reaction mechanisms). Particularly, the ESI-MS technique has been used to perform protein folding analyses and on-line kinetic studies (Zechel et al., 1998). In the latter case, the reaction kinetic can be monitored by direct injection of the reaction mixture to the ESI source, and the relative concentration of the different species can be analyzed as a function of time (Konermann and Douglas, 2002; Li et al., 2003; Wilson and Konermann, 2004). With ESI-MS it is also possible to monitor changes in protein conformation as well as protein binding or dissociation to a specific ligand as a function of time (Konermann, 1999). For instance, this technique was used in the study of the reconstitution of acid-denaturated holomyoglobin and its noncovalent binding to a hemo group (Lee et al., 1999). Likewise, the same technique has been used to study different kinetic states of an enzymatic reaction, as well as the presence of labile intermediates in enzymatic reactions that take place in the order of milliseconds (Roberts et al., 2010).

5.4 Nuclear Magnetic Resonance

The NMR is a physical phenomenon based on the quantum-mechanical properties of the atomic nucleus (Macomber, 1998). NMR-based spectroscopy is a research analytical tool that provides a great volume of data with high specificity such as structural and stereochemical information of a compound (Markwick et al., 2008). An important feature of this technique is the fact that data acquisition is relatively fast. In contrast to MS, NMR is a nondestructive technique that has applications in all areas of chemistry and some biological studies (particularly, to investigate the properties of biomolecules).

NMR spectroscopy is based on the radiofrequency energy absorption through a magnetically active nucleus which is oriented within a magnetic field. The energy absorbed causes this orientation to change. Among the atomic species which are most widely used in this spectroscopy are the following magnetically active nucleuses: proton (^{1}H), carbon (^{13}C), nitrogen (^{15}N), phosphorus (^{31}P), and fluorine (^{19}F). Samples are generally dissolved in deuterated solvents (free of protium atoms). Most common NMR spectra are representations of radiation absorption intensity versus frequency of resonance. The spectral analysis obtained shows signals whose position, shape, and size are closely related to the molecular structure of the analyzed species. The detailed study of these spectra provides valuable structural and stereochemical information. This technique also allows the analysis of samples which are not in solution. NMR in solid state is a suitable technique increasingly used for the study of structural properties of a wide range of amorphous or little crystalline materials that cannot be studied by diffraction techniques. Unlike samples in solution, spectra of samples in solid state show signals which slightly differ from the ideal spectra. However, these spectra contain unique information about the structure and dynamics of the studied material.

5.5 Applications of Nuclear Magnetic Resonance

Although the use of ^{1}H NMR to determine the metabolic fingerprint in the biomedical field is well established, its application to the field of biology and physiology of photosynthetic organisms is less extensive (Weljie et al., 2006; Ward et al., 2007). Recent studies have established NMR for high-throughput comparative analysis of plant extracts, providing metabolomic data for various applications, such as in functional genomics, or to differentiate plants from different origins or after different treatments. The success of these analyses depends on a good experimental design, since metabolite levels are dependent on the tissue (or organ) physiological stage and environmental conditions (Kim et al., 2011). The advantages of NMR over MS when applied to metabolomic studies include: relatively easy preparations of the sample, nondestructive analyses, great potential to identify a wide range of compounds, larger capacity for the conclusive identification of compounds, and structural information about unknown compounds. An important limitation of this NMR spectroscopy application is the requirement of large amounts of samples due to its low sensitivity, although recent advances have lessened this problem. In addition, the analysis of specific compounds from plant crude extracts by NMR spectra is hampered by several problems that include the spectral complexity, the fact that resonance peaks overlap with each other, and the lack of a complete spectral library of standard compounds from these organisms. However, numerous reports on the use of metabolic fingerprints in plants based on NMR to catalog the overall changes in the metabolome have been published in this last decade (Krishnan et al., 2005; Zulak et al., 2008; Kim et al., 2011; Mahrous and Farag, 2015).

Further Reading and References

De Hoffmann, E., Sroobant, V., 2007. Mass Spectrometry: Principles and Applications. third ed. J. Wiley & Sons, England.

Herbert, C.G., Johnstone, R.A.W., 2003. Mass Spectrometry Basics. CRC Press, Boca Raton, FL.

Jürgen, H.G., 2004. Mass Spectrometry. A Textbook. Springer-Verlag, Berlin, Heidelberg.

Kim, H.K., Choi, Y.H., Verpoorte, R., 2011. NMR-based plant metabolomics: where do we stand, where do we go? Trends Biotechnol. 29, 267–275.

Konermann, L., 1999. Monitoring reaction kinetics by continuous-flow methods: the effects of convection and molecular diffusion under laminar flow conditions. J. Phys. Chem. 103, 7210–7216.

Konermann, L., Douglas, D.J., 2002. Pre-steady state kinetics of enzymatic reaction studied by electrospray mass spectrometry with on-line rapid-mixing techniques. In: Neufeld, E.F., Ginsburg, V. (Eds.), Methods Enzymology, vol. 354, Academic Press, San Diego, CA, pp. 50–64.

Krishnan, P., Kruger, N.J., Ratcliffe, R.G., 2005. Metabolite fingerprinting and profiling in plants using NMR. J. Exp. Bot. 56, 255–265.

Lee, V.W.S., Chen, Y.L., Konermann, L., 1999. Reconstitution of acid-denatured holomyoglobin studied by time-resolved electrospray ionization mass spectrometry. Anal. Chem. 71, 4154–4159.

Li, Z., Sau, A.K., Shen, S., Whitehouse, C., Baasov, T., Anderson, K.S., 2003. A snapshot of enzyme catalysis using electrospray ionization mass spectrometry. J. Am. Chem. Soc. 125, 9938–9939.

Macomber, R.S., 1998. A Complete Introduction to Modern NMR Spectroscopy. John Wiley & Sons, Inc., New York.

Mahrous, E.A., Farag, M.A., 2015. Two dimensional NMR spectroscopic approaches for exploring plant metabolome: a review. J. Adv. Res. 6, 3–15.

Markwick, P.R.L., Mallavin, T., Nilges, M., 2008. Structural biology by NMR: structure, dynamics, and interactions. PLoS Comput. Biol. 4, e1000168. Available from: http://dx.doi.org/10.1371/journal.pcbi.1000168.

Roberts, A., Furdui, C., Anderson, K.S., 2010. Observation of a chemically labile, noncovalent enzyme intermediate in the reaction of metal-dependent *Aquifex pyrophilus* KD08PS by time-resolved mass spectrometry. Rapid Commun. Mass Spectrom. 24, 1919–1924.

Rodrigues, J.A., Taylor, A.M., Sumpton, D.P., Reynolds, J.C., Pickford, R., Thomas-Oates, J., 2007. Mass spectrometry of carbohydrates: newer aspects. Adv. Carbohydr. Chem. Biochem. 61, 59–141.

Ward, J.L., Baker, J.M., Beale, M.H., 2007. Recent applications of NMR spectroscopy in plant metabolomics. FEBS J. 274, 1126–1131.

Watson, J.T., Sparkman, O.D., 2007. Introduction to Mass Spectrometry: Instrumentation, Applications, and Strategies for Data Interpretation. fourth ed. John Wiley & Sons, Hoboken, NJ, 862 p.

Weljie, A.M., Newton, J., Mercier, P., Carlson, E., Slupsky, C.M., 2006. Targeted profiling: quantitative analysis of ^{1}H NMR metabolomics data. Anal. Chem. 78, 4430–4442.

Wilson, D.J., Konermann, L., 2004. Mechanistic studies on enzymatic reactions by ESI-MS using a capillary mixer with adjustable reaction chamber volume for time-resolved measurement. Anal. Chem. 76, 2537–2543.

Zechel, D.L., Konerman, L., Withers, S.G., Douglas, D.J., 1998. Pre-steady state kinetic analysis of an enzymatic reaction monitored by time-resolved electrospray ionization mass spectrometry. Biochemistry. 37, 7664–7669.

Zulak, K.G., Weljie, A.M., Vogel, H.J., Facchini, P.J., 2008. Quantitative ^{1}H NMR metabolomics reveals extensive metabolic reprogramming of primary and secondary metabolism in elicitor-treated opium poppy cell cultures. BMC Plant Biol. 8 (1), 5.

PART III

Case Studies

CHAPTER 6

Case Study: Sucrose

Chapter Outline

6.1 Introduction

Sucrose (α-glucopyranosyl-β-D-fructofuranoside) is mainly synthesized in oxygenic photosynthetic organisms (such as cyanobacteria, green algae, and land plants), as part of the carbon dioxide assimilation pathway (Salerno and Curatti, 2003; Kolman et al., 2015). The central role of sucrose in plant life can be compared with that of glucose in the animal world. It is the major product of photosynthesis and the predominant molecule of carbon transport in plants, and the principal form of carbon storage that provides a ready source of glucose and fructose for synthesis and energy (Pontis, 1977). It also plays a central role in responses to environmental stresses and in signal transduction (Tognetti et al., 2013; Ruan, 2014).

The importance of sucrose in nature is enhanced when it is taken into account that nearly all carbon compounds in the nonphotosynthetic tissues of plants derive from it. Moreover, the majority of carbon compounds present in animals come from the sucrose

Methods for Analysis of Carbohydrate Metabolism in Photosynthetic Organisms.
DOI: http://dx.doi.org/10.1016/B978-0-12-803396-8.00006-5

carbon skeleton, as animals are not autotrophic organisms and depend on plants for their subsistence and development.

The reasons for the emergence and ubiquity of sucrose in oxygenic photosynthetic organisms are still an enigma. However, there is no doubt that the properties of the molecule could have been crucial for the appearance and success of this disaccharide in nature (Salerno and Curatti, 2003). It is a nonreducing sugar that is easily hydrolyzed by diluted acid, and reasonably stable in the presence of strong alkalis. In photosynthetic organisms, sucrose can be hydrolyzed to glucose and fructose by acid or alkaline/neutral invertases (Takewaki et al., 1980; Vargas and Salerno, 2010).

The ease with which sucrose hydrolyzes comes from the nature of the glycosidic linkage (Fig. 6.1). Sucrose has the glucopyranosyl moiety in the chair (C1) conformation, and the β-D-fructofuranosyl possesses the envelope (E4) conformation. Chemically, this conformation contributes to a greater reactivity of the primary hydroxyls, which are in a less hindered or more exposed position. Another sucrose characteristic is its β-fructofuranoside form, which is a rarity, as pointed out by Edelman (1971). This structure appears in other carbohydrates, but all of them are based on sucrose and could be considered its derivatives.

The furanose configuration of the fructoside moiety of sucrose bestows the glycosidic linkage upon a very high free-energy of hydrolysis (ΔG°). Its value has been estimated as -7000 cal mol^{-1} (Hassid and Doudoroff, 1950) from the equilibrium constant ($K_{eq} = 0.053$, at pH 6.6) of the reaction catalyzed by the enzyme sucrose phosphorylase (α-D-glucopyranosyl-phosphate + D-fructose ↔ sucrose + Pi).

The value of sucrose hydrolysis energy differs from those of trehalose (-5700 cal mol^{-1}), maltose (-3000 cal mol^{-1}), and lactose (-3000 cal mol^{-1}), but it can be compared with those of UDP-glucose (about -7600 to -8000 cal mol^{-1}) and ATP (-6900 cal mol^{-1}) (Leloir et al., 1960). The relatively high value for the sucrose glycosidic bond has led to suggest that this may account for the distinctive biological function of sucrose in plants (Hassid, 1951). On the contrary, Arnold (1968), who studied the role of sucrose in plant

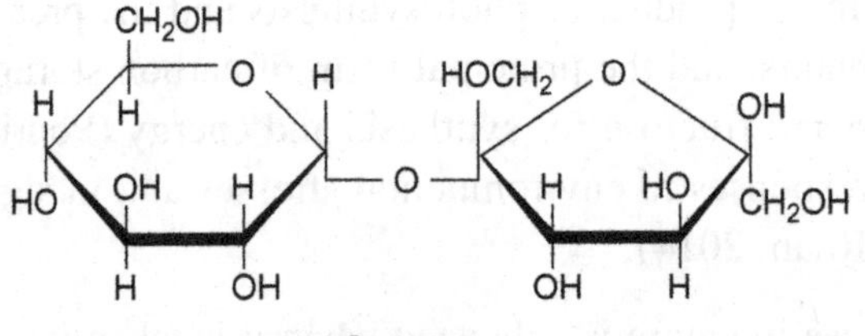

Sucrose

Figure 6.1
Sucrose structure (α-D-glucopyranosyl-β-D-fructofuranoside).

metabolism, reported that the use of sucrose has no real energetic advantage over using glucose when considering the number of ATP molecules that sucrose can produce. In his analysis, Arnold only considered two enzymes as acting on sucrose degradation: sucrose phosphorylase and invertase, omitting sucrose synthase, which actually catalyzes a nucleotide pyrophosphorolysis of sucrose (Pontis, 1977), allowing the cell to use the energy of the glycosidic linkage for synthetic activities.

Among them, sucrose is the only disaccharide biosynthesized in nature whose glycoside linkage can be cleaved (by the action of sucrose synthase) keeping its free energy. This fact gives sucrose a special place and could explain its selection by plants as the main substance used for translocation and storage. Furthermore, sucrose is one of the sugars used as signal molecules for different metabolic processes.

6.2 Description of Enzyme Reactions

6.2.1 Biosynthesis of Sucrose

It is generally accepted that sucrose synthesis takes place in the cell cytosol of most plant tissues, as well as in green algae and unicellular and filamentous cyanobacteria. The enzyme sucrose-phosphate synthase (SPS, EC 2.4.1.14) catalyzes the first step in the biosynthesis pathway (Leloir and Cardini, 1955) in which sucrose-6-phosphate is synthesized from fructose-6-phosphate and a sugar nucleotide (UDP-glucose for land plants and unicellular algae, or mainly UDP-glucose or ADP-glucose for cyanobacteria) (Winter and Huber, 2000; Kolman et al., 2015). In a second step, the disaccharide-phosphate is dephosphorylated by sucrose-phosphate phosphatase (SPP, EC 3.1.3.24), a highly specific phosphatase that displaces the reversible SPS reaction from equilibrium in vivo yielding sucrose (Cardini et al., 1955). The two-step pathway can be summarized as follows, where N is mostly uracil in plants and unicellular algae, and adenine or uracil in cyanobacteria:

$$\text{NDP-glucose} + \text{Fructose-6-phosphate} \xrightarrow{\text{SPS}} \text{NDP} + \text{Sucrose-6-phosphate}$$

$$\text{Sucrose-6-phosphate} \xrightarrow{\text{SPP}} \text{Sucrose} + \text{Orthophosphate (Pi)}$$

6.2.2 Sucrose Cleavage and Hydrolysis

The sucrose glucosydic linkage can be broken by at least two different enzymatic activities that lead either to its reversible cleavage (sucrose synthase) or to its irreversible hydrolysis to hexoses (invertases).

Sucrose synthase (SuS, EC 2.4.1.13) is a glucosyltransferase that catalyzes a reversible cleavage of sucrose (Cardini et al., 1955) in the presence of a nucleoside diphosphate (Winter and Huber, 2000; Kolman et al., 2015).

$$\text{Sucrose} + \text{UDP} \xleftrightarrow{\text{SuS}} \text{UDP-glucose} + \text{Fructose}$$

UDP can be replaced by ADP, GDP, CDP, or TDP; however, UDP is the preferred substrate in plants, and ADP in cyanobacteria (Pontis, 1977; Salerno and Curatti, 2003). Even though SuS may synthesize sucrose in vivo (Martinez-Noël and Pontis, 2000), it mostly acts in vivo in the cleavage of sucrose, playing an important role in the mobilization of the stored sucrose and in the synthesis of nucleoside diphosphate sugar (Avigad, 1990). The sucrose cleavage reaction allows conserving the energy of the sucrose glycoside linkage in the sugar–nucleotide products, which in turn, are precursors for the synthesis of other compounds, such as cellulose, callose, or starch in plants, or glycogen in cyanobacteria. SuS is ubiquitous in plants and particularly active in plant sink tissues, such as roots, young leaves, developing seeds, or tubers (Winter and Huber, 2000; Koch, 2004). On the contrary, in cyanobacteria SuS is not widespread, and it can only be found in filamentous nitrogen-fixing strains and in a few unicellular strains (Kolman et al., 2015).

Invertases catalyze sucrose hydrolysis to glucose and fructose providing monosaccharides to the glycolytic pathway, or generating glucose-mediated signals to regulate plant cell metabolism.

$$\text{Sucrose} \xrightarrow{\text{Invertase}} \text{Glucose} + \text{Fructose}$$

Two types of invertases, initially differentiated by their optimum pH in vitro, occur in plants and algae: (1) acid invertases (EC 3.2.1.26, β-D-fructofuranosidases) with optimum pH about 5, which are found in the cell wall and in the vacuole and (2) alkaline/neutral invertases (α-glucosidases) with optimum pH between 6.5 and 8.0, which are localized not only in the cytosol but also in organelles (Vargas and Salerno, 2010). The alkaline/neutral type is the only invertase isoform present in most cyanobacteria. Recently, it was reported in the cyanobacterium *Synechocystis* sp. PCC 7002 that sucrose is hydrolyzed to hexoses by the enzyme amylosucrase (AMS) (EC 2.4.1.4), a glycoside hydrolase only reported in bacteria (Perez-Cenci and Salerno, 2014). This enzyme is also able to transfer the glucose moiety to a soluble maltooligosaccharide or to an insoluble α-(1→4)-glucan (amylose-like polymers). Cyanobacterial AMS might have been acquired by lateral gene transfer from bacteria. Interestingly, sucrose phosphorylase (EC 2.4.1.7), present in bacteria, has not been found either in plants or in cyanobacteria (Kolman et al., 2015).

Experimental Protocols

6.3 Sucrose Extraction

The procedure for sucrose extraction is very similar for different plant tissues or cells. Sucrose is extracted together with other soluble sugars with lightly alkaline water.
The initial step may differ according to the material selected. In the case of aerial parts (leaves or shoots), fresh or freeze-dried weighed material is ground in a mortar under liquid nitrogen and reduced to powder. Tubers or roots must be cut into small pieces before grinding. Cell pellets of green algae or cyanobacteria can also be ground under liquid nitrogen. The powdered material is carefully suspended in alkaline water (brought to pH ~8 with ammonia solution) according to a ratio of 2 mL of water per gram of fresh weight. The suspension is heated at 100°C for 5 min under stirring with a glass rod and centrifuged at 12,000 × g for 5 min at 4°C. The supernatant is transferred to another tube. The pellet extraction is repeated twice more and the supernatants are combined and freeze-dried. Monosaccharides present in the sample can be destroyed by heating 10 min in alkali 0.4 N. Sucrose is estimated by either of the methods described in Chapter 1, Determination of Carbohydrates Metabolism Molecules.

6.4 Enzyme Activity Assays

Protocols for enzyme activity determinations depend on the source of the enzyme to be assayed and its purification degree. In crude extracts or homogenates, special consideration should be given for enzyme stability and the presence of other accompanying activities or compounds that could interfere in the assay (see Chapter 2, Preparation of Protein Extracts). For routine assays, enzyme activities are measured at saturating substrate levels at an appropriate pH. Different compounds are usually added to the reaction mixture (such as divalent cations, unspecific phosphatase inhibitor such as NaF, arbutin), which will be discussed for each enzyme in particular.

SPS and SuS (in the sucrose synthesis direction) activities are assayed by quantifying any of the two reaction products (sucrose-6-phosphate/sucrose or UDP/ADP). Usually, colorimetric or spectrophotometric methods are appropriate for determining activities of partially purified enzymes or for kinetic studies with highly purified proteins. Depending on the enzyme source these methods are not suitable for measuring enzyme activities in homogenates. The use of a labeled substrate (UDP-[^{14}C]-glucose or other sugar nucleotide) followed by the separation of the product formed by passing the reaction mixture through anionic resins resulted in the successful measurement of SPS and SuS activities in crude extracts from plants, algae, and cyanobacteria. This procedure is particularly chosen when the amount of product formed is lower than the detection limit of conventional methods, or when crude extracts contain molecules that can interfere either with the analytical method or react with substrates or products of the reaction to be quantified. For example, SPS activity determination can be overestimated in a crude extract containing SuS and unspecific phosphatases, which generates fructose from fructose-6-phosphate. Routinely, colorimetric methods are used for sucrose (or sucrose-6-phosphate) determination and it is necessary to remove unreacted substrates (fructose or fructose-6-phosphate) or monosaccharides from the reaction mixture. This can be achieved by destroying the monosaccharides with hot alkali or by reducing them to alcohol with sodium borohydride. Once this objective is accomplished, the sucrose formed can be determined by either of the methods described in Chapter 1, Determination of Carbohydrates Metabolism Molecules. Also determination of the other product formed in the SPS or SuS reaction (UDP or ADP) can be carried out after incubation with pyruvate kinase (as auxiliary enzyme) followed by the addition of dinitrophenylhydrazine or lactic dehydrogenase (see Chapter 1, Determination of Carbohydrates Metabolism Molecules, Section 1.7.2).

Sucrose cleavage reaction catalyzed by SuS can be assayed by measuring the amount of sugar nucleotide formed (UDP-glucose or ADP-glucose) or the appearance of fructose. UDP-glucose is quantified by measuring UDP-glucose dehydrogenase activity, while ADP-glucose is estimated after reacting with ADP-glucose pyrophosphorylase coupled with phosphoglucomutase and glucose-6-phosphate dehydrogenase/NADP (see Chapter 1, Determination of Carbohydrates Metabolism Molecules). Fructose is determined after incubation with auxiliary coupled enzymes (hexokinase, phosphoglucose isomerase, and glucose-6-phosphate dehydrogenase/$NADP^+$) (see Chapter 1, Determination of Carbohydrates Metabolism Molecules). The sensitivity of both methods can be considerably increased using fluorometric techniques (Jones et al., 1977) described in Chapter 1, Determination of Carbohydrates Metabolism Molecules.

Sucrose hydrolysis catalyzed by invertase or by amylosucrase can be quantified by measuring the amount of glucose and fructose. Colorimetric methods or enzymatic coupled reactions involving hexokinase, phosphoglucose isomerase, and glucose-6-phosphate dehydrogenase/$NADP^+$ are routinely employed (see Chapter 1, Determination of Carbohydrates Metabolism Molecules).

6.4.1 Determination of SuS (Sucrose Synthesis Direction) or SPS Activity

6.4.1.1 Activity assay using labeled substrates

Principle

SuS (in the sucrose synthesis direction) and SPS can be assayed in protein extracts using radioactive UDP-glucose. The resulting labeled product (sucrose or sucrose-6-phosphate, SuS or SPS reaction product, respectively) is separated through an anionic exchange resin column. The procedure differs according to the enzyme that is assayed (Salerno et al., 1979).

Reagents

UDP-[^{14}C]-glucose	50 mM (160–240.000 cpm μmol^{-1})
Fructose (for SuS) or fructose-6-phosphate (for SPS)	100 mM
NaF (for SPS)	200 mM
$MgCl_2$	200 mM
Tris-HCl buffer, pH 7.8–8.0 (for SuS) or Hepes-NaOH buffer, pH 7.0 (for SPS)	1 M
Glycine-NaOH buffer pH 10.0	1 M
Anionic resin AG1-X4 Cl^- (200–400 mesh)	
Alkaline phosphatase, free of phosphodiesterase activity	

Procedures

(A) SPS assay

In a total volume of 50 μL, mix 0.5 μmol UDP-[^{14}C] glucose (specific activity 160,000–240,000 cpm · μmol^{-1}), 0.5 μmol fructose-6-phosphate, 1 μmol NaF, 5 μmol Hepes-NaOH buffer (pH 7.0), and an aliquot of the enzyme preparation. Reaction blank tubes contain the same mixture but omit fructose-6-phosphate. Incubate the tubes for 10–20 min at 30°C. Stop the reaction by adding 10 μL of glycine-NaOH buffer 1 M (pH 10), and heat at 100°C for 1 min. After cooling down, complete the blanks with fructose-6-phosphate. Add 5 μL of alkaline phosphatase (10 U · mg^{-1} protein). Incubate for 20 min at 37°C and stop the reaction by adding 0.2 mL of cool water (ice temperature). Load the complete mixture on an AG1-X4 Cl^- (200–400 mesh) column (0.6 × 2 cm). The remaining UDP-[^{14}C]-glucose is retained in the resin. Elute the [^{14}C]-sucrose (released from [^{14}C]-sucrose-6-phosphate by the action of the alkaline phosphatase) with 2 mL of water. Collect the eluates in vials. Add scintillation liquid and quantify radioactivity in a scintillation spectrometer.

Comments

Alkaline phosphatase should be free of phosphodiesterase since this activity can hydrolyze labeled UDP-glucose, with the concomitant production of labeled glucose-1-phosphate. If this occurs, when alkaline phosphatase is added, it could also hydrolyze this phosphoric ester. This could increase the reaction blanks, since [^{14}C]-glucose will accompany the [^{14}C]-sucrose produced in the reaction. This can be avoided if after incubation with alkaline phosphatase, a mixture of ATP, $MgCl_2$, NADP, hexokinase, and glucose-6-phosphate dehydrogenase is added. The amount of ATP and NADP must exceed the total amount of UDP-glucose and fructose-6-phosphate.

In a crude extract, where SPS and SuS are simultaneously present, SuS activity can be inhibited by adding to the incubation mixture 10 mM arbutin or 10 mM phenyl-glucoside. Both substances produce a 95% inhibition of SuS activity, affecting SPS activity by only 10%.

A classical scintillation liquid used for aqueous samples consists of PPO-POPOP:Triton (2.5:1), prepared dissolving 4 g of PPO and 0.3 g of POPOP in one liter of toluene. However, this cocktail contains hazardous solvents whose storage in laboratories is restricted and should be used in fume hoods. Nowadays safer scintillation liquids can be purchased.

(B) SuS assay

In a total volume of 50 μL, mix 0.5 μmol UDP-[^{14}C]-glucose (specific activity 160,000–240,000 cpm.μmol^{-1}), 0.5 μmol fructose, 5 μmol Tris-HCl buffer (pH 7.8–8.0), and an aliquot of the enzyme preparation. Reaction blank tubes contain the same mixture except for fructose. Incubate the tubes for 10–20 min at 30°C. Stop the reaction by heating at 100°C for 1 min. After cooling down, complete the blank tubes with fructose. Add 0.2 mL of cool water (ice temperature). Load the complete mixture on an AG1-X4 Cl^- (200–400 mesh) column (0.6 × 2 cm). The remaining UDP-[^{14}C] glucose is retained in the resin. Elute the [^{14}C]-sucrose with 2 mL of water. Collect the eluates in vials. Add scintillation liquid and quantify [^{14}C]-sucrose in a scintillation spectrometer.

6.4.1.2 Activity assay using nonlabeled substrates

Reagents

UDP-glucose	100 mM
Fructose (for SuS) or fructose-6-phosphate (for SPS)	100 mM
NaF (for SPS)	200 mM
$MgCl_2$	200 mM
Tris-HCl buffer, pH 7.8–8.0 (for SuS) or Hepes-NaOH buffer, pH 7.0 (for SPS)	1 M

Procedure

In general, to assay SuS activity (in the sucrose synthesis direction) or SPS activity, the reaction mixture contains in a 50-μL total volume, 0.5 μmol of UDP-glucose (or ADP-glucose for cyanobacterial SuS or SPS assays), 0.5 μmol of fructose (for SuS) or 0.5 μmol of fructose-6-phosphate (for SPS), 1 μmol of NaF (for SPS), 10 μmol of Tris-HCl buffer (pH 7.8–8.0) (for SuS) or Hepes-NaOH buffer (pH 7.0) (for SPS), and an aliquot of the enzyme preparation. Reaction blank tubes contain the same mixture except for the sugar nucleotide. The mixture is usually incubated at 30°C for 10–20 min. Often $MgCl_2$ at 10–20 mM final concentration is added to the incubation mixture to increase activity. Glucose-6-phosphate (at a concentration fivefold higher than that of fructose-6-phosphate) is added to the SPS assay in leaves (Salvucci et al., 1990). In accordance with the analytical method that will be used to quantify product formation, the reaction is stopped in different ways, as described below, and blanks are completed.

6.4.1.2.1 Measurement of products by colorimetric methods

Principle

Sucrose-6-phosphate (for SPS activity determination) or sucrose (for SuS activity determination) are estimated by a colorimetric method that quantifies the fructose moiety. The presence of substrate excess (fructose-6-phosphate or fructose) in the incubation mixture interferes, and consequently these unreacted substrates should be destroyed by either of the two approaches described below.

(A) The quantification of the reaction product (sucrose or sucrose-6-phosphate) is carried out after destroying the remaining substrate (fructose or fructose-6-phosphate) by heating with alkali (Cardini et al., 1955), following the thiobarbituric acid method (see Chapter 1, Determination of Carbohydrates Metabolism Molecules).

Reagents

NaOH	0.5 N
Thiobarbituric acid Reagent (TBA-HCl)	

Procedure

After incubation of the reaction mixture (see Section 6.4.1.2) to measure SPS or SuS activity, stop the reaction by adding 200 μL of 0.5 N NaOH. Enzyme reaction blanks do not contain UDP-glucose (or ADP-glucose for cyanobacterial SuS or SPS), which is added after the alkali addition. After vortex stirring, heat the mixtures at 100°C for 10 min. Cool to room temperature, add 600 μL of the TBA-HCl reagent, and heat at 100°C for 7 min. After cooling, read absorbance at 432 nm. The difference in absorbance value between the complete mixture and its corresponding reaction blank yields to the amount of sucrose (for SuS) or sucrose-6-phosphate (for SPS) formed. This method allows the detection of 10–120 μmoles of product.

Comments

SPS and SuS from cyanobacteria can accept other nucleoside-diphosphate sugars, such as ADP-glucose (Porchia and Salerno, 1996; Porchia et al., 1999). Sodium fluoride is only added to measure SPS activity to inhibit unspecific phosphatase activities that can be present in partially purified protein preparations, and that could interfere in SPS activity determination by consuming the substrate (fructose-6-phosphate). Also, SPS activity could be overestimated in the case where SuS is an accompanying protein. Thus, fructose molecules liberated from fructose-6-phosphate by unspecific phosphatases can be a substrate in the SuS reaction which can synthesize sucrose.

(B) The quantification of the reaction product (sucrose or sucrose-6-phosphate) is carried out after reducing the remaining substrate (fructose or fructose-6-phosphate) to alcohol with borohydride (Rorem et al., 1960). Sucrose or sucrose-6-phosphate are determined by the TBA method (see Chapter 1, Determination of Carbohydrates Metabolism Molecules).

Reagents

$NaOH$-$NaBH_4$ solution	0.025 N NaOH containing 37.5 $mg \cdot mL^{-1}$ of $NaBH_4$
Glacial acetic acid	

Procedure

After incubation of the mixture to measure SPS or SuS activity, stop the reaction by heating 1 min at 100°C. Add 400 μL of $NaOH$-$NaBH_4$ solution to each tube. Incubate at 30°C for 1 h and heat at 100°C for 5 min to complete reduction of sugars to alcohol. After cooling down, add 50 μL of acetic acid to destroy the excesses of borohydride. Finally, quantify sucrose or sucrose-6-phosphate by the Thiobarbituric acid method.

6.4.1.2.2 Measurement of sucrose after hydrolysis

Principle

The production of sucrose by the action of SuS can be quantified after the disaccharide hydrolysis by the addition of exogenous invertase (β-fructofuranosidase) to the incubation mixture. The resulting hexoses (glucose and fructose) can be estimated by either of the methods described in Chapter 1, Determination of Carbohydrates Metabolism Molecules. An aliquot of SuS preparation is incubated with SuS substrates (see Section 6.4.1.2). The reaction is stopped by boiling for 1 min and sucrose hydrolysis to hexoses is described below.

Reagents

Acetate buffer (pH 4.5)	1 M
Invertase from baker's yeast	
Tris-HCl buffer (pH 8.0)	100 mM
Bovine serum albumin (BSA)	0.2%

Invertase solution: Dissolve 8 mg of yeast invertase in 250 μL of 100 mM Tris-HCl (pH 8.0) buffer. Add 100 μL of 0.2% BSA and 650 μL of distilled water. Usually a 1/10 dilution of this preparation is used for hydrolyzing sucrose in SuS assays; however it is recommended to check each invertase preparation and adjust the amount before use, by performing a time course of the reaction with a standard sucrose solution.

Procedure

Once the SuS enzymatic reaction is stopped (50-μL total volume, see Section 6.4.1.2), add 5 μL of 1 M acetate buffer (pH 4.5) and 5 μL of yeast invertase (appropriately diluted).

Prepare a blank tube without adding invertase. Incubate the tubes at 37°C for 30 min. Hexoses formed can be determined by a colorimetric method or by an enzymatic assay, described below.

(A) Determination of hexoses by the Somogyi–Nelson method

Stop the invertase reaction adding 500 μL of the diluted Somogyi's reagent and follow the procedure indicated in Chapter 1, Determination of Carbohydrates Metabolism Molecules, Section 1.2. Sucrose produced in the SuS reaction is calculated as follows:

$$\text{Sucrose } (\mu g) = (\text{Total reducing sugars } (\mu g) - \text{Free reducing sugars } (\mu g))/2$$

where "Total reducing sugars" correspond to the value obtained from the complete incubation including invertase and "Free reducing sugars" correspond to the value obtained from the incomplete incubation without invertase (blank).

(B) Determination of hexoses by coupled enzyme reactions

Stop the invertase reaction by heating for 1 min at 100°C and follow the procedure describe in Chapter 1, Determination of Carbohydrates Metabolism Molecules, Section 1.6.

6.4.1.2.3 Measurement of UDP

Principle

SuS (in the sucrose synthesis direction) or SPS activity can be estimated by measuring UDP formation. After enzymatic incubation, the reaction is stopped by heating for 1 min at 100°C, and UDP is quantified after its conversion to pyruvate by the addition of pyruvate kinase to the reaction mixture. Pyruvate kinase catalyzes the phosphorylation of nucleoside diphosphates to nucleoside triphosphates with phosphoenolpyruvate, with the formation of pyruvate, which can be estimated either by a colorimetric reaction (eg, with 2,4-dinitrophenylhydrazone) or as NADH oxidation after incubation with lactic dehydrogenase. The complete procedure is described in Chapter 1, Determination of Carbohydrates Metabolism Molecules, Section 1.16.

Procedures

Once the enzymatic reaction (50-μL total volume) is stopped by boiling for 1 min, follow the colorimetric method procedure using 2,4-dinitrophenylhydrazine (see Chapter 1, Determination of Carbohydrates Metabolism Molecules, Section 1.16.1) or remove denatured protein by centrifugation at 13,000 × g for 10 min and determine the amount of UDP formed according to the assay system coupled to the NADH oxidation procedure (see Chapter 1, Determination of Carbohydrates Metabolism Molecules, Section 1.16.2).

6.4.2 Determination of SuS Activity (in the Sucrose Cleavage Direction)

Principle

SuS (in the sucrose cleavage direction) can be estimated by measuring fructose formation. After incubation of the enzymatic preparation with the substrates, fructose formed can be quantified either by a colorimetric method, such as Somogyi–Nelson (see Chapter 1, Determination of Carbohydrates Metabolism Molecules, Section 1.2), or by an enzymatic assay (see Chapter 1, Determination of Carbohydrates Metabolism Molecules, Section 1.5)

Reagents

Sucrose	1 M
UDP	50 mM
Hepes-NaOH buffer (pH 6.5)	1 M

Procedure

In general, to assay SuS activity (in the sucrose cleavage direction) the reaction mixture contains, in a total volume of 50 μL, 10 μmol of sucrose, 0.25 μmol of UDP, 5 μmol of Hepes-NaOH buffer (pH 6.5), and an aliquot of the enzyme preparation. The mixture is usually incubated at 30°C for 10–20 min. Prepare blanks without UDP and/or enzyme addition. Stop the reaction according to the method used for quantifying fructose. Complete blanks after stopping the reaction.

(A) Determination of fructose by the Somogyi–Nelson method

Stop the reaction adding 500 μL of the diluted Somogyi's reagent and follow the procedure indicated in Chapter 1, Determination of Carbohydrates Metabolism Molecules, Section 1.2. Activity is given by the amount of fructose formed, per unit of time and enzyme.

(B) Determination of fructose by coupled enzyme reactions

Stop the SuS enzymatic reaction by heating for 1 min at 100°C and follow the procedure described in Chapter 1, Determination of Carbohydrates Metabolism Molecules, Section 1.6.

6.4.3 Determination of SPP Activity

Principle

The method generally used is based on the determination of the inorganic phosphate (Pi) released by SPP action. The reaction is highly specific for sucrose-6-phosphate and has an absolute requirement of magnesium ion, which allows distinguishing SPP from other unspecific phosphatases. The inorganic phosphate determination is performed by the Chifflet's method (see Chapter 1, Determination of Carbohydrates Metabolism Molecules).

Reagents

Sucrose-6-phosphate	10 mM
$MgCl_2$	100 mM
Hepes-NaOH buffer (pH 7.0)	1 M

Procedure

In a total volume of 50 μL, the SPP reaction mixture contains: 0.05 μmol of sucrose-6-phosphate, 0.5 μmol of $MgCl_2$, 5 μmol of Hepes-NaOH buffer (pH 6.5–7.0), and an aliquot of the enzyme preparation. Incubate the mixture at 30°C for 10–20 min. Prepare blanks without sucrose-6-phosphate or enzyme addition. Stop the reaction by addition of 100 μL of 5 N sulfuric acid, complete blanks, and continue with the protocol of the Fiske–Subbarow assay.

6.4.4 Determination of Invertase and Amylosucrase Activity

Principle

Sucrose hydrolysis by invertases (acid or alkaline/neutral isoenzymes) and by amylosucrase is determined by quantifying glucose and fructose. Incubation mixture pHs vary according to the enzyme to be assayed: pH between 4.5 and 5.0, for the acid invertase assay, pH between 6.5 and 8.0, for the alkaline/neutral invertase assay, and pH 7.5, for the cyanobacterial amylosucrose assay. After incubation, the hexoses formed can be quantified either by a colorimetric method (see Chapter 1, Determination of Carbohydrates Metabolism Molecules, Section 1.2) or by an enzymatic assay (see Chapter 1, Determination of Carbohydrates Metabolism Molecules, Section 1.5).

Reagents

Sucrose	1 M
Sodium acetate buffer (pH 4.8) (acid invertase)	0.5 M
Potassium phosphate buffer (pH 7.5) (alkaline/neutral invertase)	1 M

Procedure

In a total volume of 50 μL, the invertase or amylosucrase reaction mixture contains: 10 μmol of sucrose, 5 μmol of the appropriate buffer, and an aliquot of the enzyme preparation. Incubate the mixture at 30°C for 10–20 min. Prepare blanks without sucrose or enzyme addition. Stop the reaction according to the method used for quantifying fructose. Complete blanks after stopping the reaction. Proceed as indicated in Section 6.4.2.

Comments

Most filamentous heterocyst-forming cyanobacteria have two alkaline/neutral invertases (named A and B) (Vargas et al., 2003). The assay for the neutral isoform A is carried out at pH 6.8 and at pH 8.0 (in the presence of 20 mM $MgCl_2$) for the alkaline isoform B.

To determine amylosucrase activities, hexoses are quantified by coupling hexokinase, phosphoglucose isomerase plus glucose 6-phosphate dehydrogenase, in the presence of NADP (see Chapter 1, Determination of Carbohydrates Metabolism Molecules, Section 1.6). Glucose and fructose can be distinguished by adding or omitting phosphoglucose isomerase, respectively. The production of fructose corresponds to the total sucrose consumed (total amylosucrase activity). The amount of glucose is a measure of sucrose hydrolysis by amylosucrase activity. Transglycosylation activity is calculated by subtracting the amount of glucose from the amount of fructose released.

6.5 Reaction Stoichiometry

Stoichiometry is the determination of quantitative relationships between the reactants (substrates) and products along a chemical reaction. In the study of the stoichiometry of an enzymatic reaction, substrate and product concentrations must be determined after a certain reaction time. The decrease in substrate amounts should match the amounts of product formed.

As an example, in the case of the reaction catalyzed by SPS, the decrease of UDP-glucose and fructose-6-phosphate quantities must be equal to the amount of UDP and sucrose-6-phosphate formed. In a similar way, stoichiometry of sucrose cleavage by SuS is determined by measuring UDP-glucose (or ADP-glucose) and fructose formed from the decreased levels of sucrose and UDP (or ADP). UDP-glucose could be determined by UDP-glucose dehydrogenase, fructose-6-phosphate by phosphoglucoisomerase coupled to glucose-6-phosphate dehydrogenase, and ADP-glucose by ADP-glucose pyrophosphorylase coupled to phosphoglucomutase and glucose-6-phosphate dehydrogenase/$NADP^+$ (see Chapter 1, Determination of Carbohydrates Metabolism Molecules). The determination of UDP, sucrose and sucrose-6-phosphate formed could be determined by the methods described above.

6.6 Isolation and Characterization of Reaction Products

The products formed in the enzymatic reaction in a novel biological system must be both analytically quantified and characterized after isolation to verify their identity. In the case of sucrose formation, reaction products could be separated and identified by paper chromatography or separated by anion exchange borate column chromatography, or in an exchange high-performance liquid chromatography (HPLC) column.

6.6.1 Paper Chromatography

Isolation of sucrose formed in the reaction catalyzed by the enzyme SuS could be achieved by paper chromatography, developing the chromatogram with butanol:pyridine:water (6:4:3) solvent (see Chapter 1, Protein and Carbohydrate Separation and Purification). In this solvent, sucrose moves slower than glucose and fructose, while UDP-glucose has no mobility, remaining at the origin as the other product of UDP reaction. Preparative chromatography can be done using Whatman N°3 paper.

Isolation and separation of sucrose-6-phosphate from the other products catalyzed by the SPS reaction could be achieved by paper chromatography using ammonium-acetate 1 M (pH 3.8), ethanol 95% (3:7.5). In this solvent, both sugar nucleotide and sugar phosphates move, but the latter runs faster than the former. UDP-glucose, UDP, fructose-6-phosphate, and sucrose-6-phosphate can be separated in the following order by developing the chromatogram for 20 h: sucrose-6-phosphate (the fastest compound), fructose-6-phosphate, UDP-glucose, and UDP (the slowest compound). If a free sugar glucose or fructose is formed during the reaction, these monosaccharides move with the solvent front.

6.6.2 Separation by Anion Exchange Borate Column

Separation of sugars and sugar phosphates by chromatography through ion exchange columns can be achieved by generating a sugar–borate complex, which confers a negative charge to sugars allowing their separation (see Chapter 1, Case Study: Sugar Phosphates). Consequently, an analytical or preparative anionic borate column can be successfully used. In these columns, sugar–borate complexes elute first, followed by sugar phosphates. Sodium chloride is added to elute sugar nucleotides that are retained in the column. For example, 110 μmoles of sucrose-6-phosphate produced from the SPS reaction mixture can be separated in a preparative column (Dowex 1 borate of 2.6×54 cm or AG1-X4 Cl^-) whose elution is achieved with three lineal gradients of ammonium borate (pH 8.6). The first gradient is produced with 150 mL ammonium borate 0.3–0.4 M, the second one with 0.4–0.6 M, and the third one with 0.6–0.7 M. Finally, a 0.7 M ammonium borate solution is passed through the column with the same pH (Salerno and Pontis, 1986; Fig. 6.2).

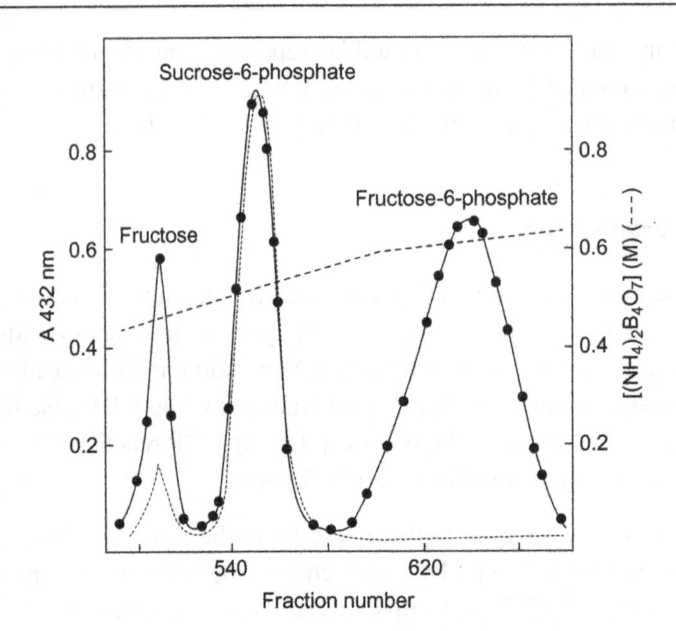

Figure 6.2
Dowex 1-borate chromatography of sucrose-phosphate synthase products. An aliquot of purified wheat germ SPS was incubated at 30°C in a reaction mixture (pH 6.5) containing fructose-6-phosphate and UDP-glucose (see Section 6.4.1.2.1(A)). The reaction was stopped and the mixture was loaded to a Dowex 1-borate column to separate substrates from products. Fractions were analyzed by the thiobarbituric acid method (see Chapter 1, Determination of Carbohydrates Metabolism Molecules) after addition of 0.4 M alkali (dotted line), or without alkali addition (full line). *Republished with permission from Salerno, G.L., Pontis, H.G., 1986. Síntesis semicontinua de sacarosa-6-fosfato. An. Asoc. Quim. Argent. 6, 719–725.*

Further Reading and References

Arnold, W.N., 1968. The selection of sucrose as the translocate in higher plants. J. Theor. Biol. 21, 13–223.

Avigad, G., 1990. Disaccharides. In: Dey, P. (Ed.), Methods in Plant Biochemistry, vol. 2. Academic Press, London, pp. 111–189.

Cardini, C.E., Leloir, L.F., Chiriboga, J., 1955. The biosynthesis of sucrose. J. Biol. Chem. 214, 149–156.

Edelman, J., 1971. In: Yudkin, J., Edelman, J., Hough, L. (Eds.), Sugar. Butterworths, London, pp. 95–102.

Hassid, W.Z., 1951. Sucrose biosynthesis. In: McElroy, W.S., Glass, B. (Eds.), Phosphorus Metabolism, vol. 1. John Hopkins Press, Baltimore MD, pp. 11–42.

Hassid, W.Z., Doudoroff, M., 1950. The biosynthesis of sucrose. Adv. Carbohydr. Chem. Biochem. 5, 29–48.

Jones, M.G.K., Outlaw Jr., W.H., Lowry, O.H., 1977. Enzymatic assay of 10^{-7} to 10^{-14} moles of sucrose in plant tissues. Plant Physiol. 60, 379–383.

Koch, K., 2004. Sucrose metabolism: regulatory mechanisms and pivotal roles in sugar sensing and plant development. Curr. Opin. Plant Biol. 7, 235–246.

Kolman, M.A., Nishi, C.N., Perez-Cenci, M., Salerno, G.L., 2015. Sucrose in cyanobacteria: from a salt-response molecule to play a key role in nitrogen fixation. Life. 5, 102–126.

Leloir, L.F., Cardini, C.E., 1955. The biosynthesis of sucrose phosphate. J. Biol. Chem. 2124, 157–161.

Leloir, L.F., Cardini, C.E., Cabib, E., 1960. Utilization of free energy for the biosynthesis of saccharides. In: Florkin, M., Mason, H.S. (Eds.), Comparative Biochemistry, vol. 21. Academic Press, New York, pp. 97–138.

Martinez-Noël, G., Pontis, H.G., 2000. Involvement of sucrose synthase in sucrose synthesis during mobilization of fructans in dormant Jerusalem articjoke tubers. Plant Sci. 159, 191–195.

Perez-Cenci, M., Salerno, G.L., 2014. Functional characterization of *Synechococcus* amylosucrase and fructokinase encoding genes discovers two novel actors on the stage of cyanobacterial sucrose metabolism. Plant Sci. 2014 (224), 95–102.

Pontis, H.G., 1977. Riddle of sucrose. In: Northcote, D.H. (Ed.), Plant Biochemsitry, vol. 2. University Park Press, Baltimore, MD, London, pp. 79–117.

Porchia, A.C., Salerno, G.L., 1996. Sucrose biosynthesis in a prokaryotic organism: presence of two sucrose-phosphate synthases in *Anabaena* with remarkable differences compared with the plant enzymes. Proc. Natl. Acad. Sci. USA 93, 13600–13604.

Porchia, A.C., Curatti, L., Salerno, G.L., 1999. Sucrose metabolism in cyanobacteria: sucrose synthase from *Anabaena* sp. strain PCC 7119 is remarkably different from the plant enzymes with respect to substrate affinity and amino-terminal sequence. Planta. 1, 34–40.

Rorem, E.S., Walker Jr., H.G., McCready, R.M., 1960. Biosynthesis of sucrose and sucrose-phosphate by sugar beet leaf extracts. Plant Physiol. 35, 269–272.

Ruan, Y.L., 2014. Sucrose metabolism: gateway to diverse carbon use and sugar signaling. Annu. Rev. Plant Biol. 65, 33–67.

Salerno, G.L., Curatti, L., 2003. Origin of sucrose metabolism in higher plants: when, how and why. Trends Plant Sci. 8, 63–69.

Salerno, G.L., Gamundi, S.S., Pontis, H.G., 1979. A procedure for the assay of sucrose synthase and sucrose phosphate synthase in plant homogenates. Anal. Biochem. 93, 196–199.

Salerno, G.L., Pontis, H.G., 1986. Síntesis semicontinua de sacarosa-6-fosfato. An. Asoc. Quim. Argent. 6, 719–725.

Salvucci, M.E., Drake, R.R., Haley, B.E., 1990. Purification and photoaffinity labeling on sucrose-phosphate synthase from spinach leaves. Arch. Biochem. Biophys. 281, 212–218.

Takewaki, S., Chiba, S., Kimura, A., Matsui, H., Koike, Y., 1980. Purification and properties of alpha-glucosidases of the honey bee *Apis mellifera* L. J. Agricult. Biol. Chem. 44, 731–740.

Tognetti, J., Pontis, H.G., Martínez-Noël, G., 2013. Sucrose signaling in plants. A world yet to be explored. Plant Signal. Behav. 8 (3), e23316.

Vargas, W.A., Salerno, G.L., 2010. The Cinderella story of sucrose hydrolysis: alkaline/neutral invertases, from cyanobacteria to unforeseen roles in plant cytosol and organelles. Plant Sci. 1, 1–8.

Vargas, W., Cumino, A., Salerno, G.L., 2003. Cyanobacterial alkaline/neutral invertases. Origin of sucrose hydrolysis in the plant cytosol? Planta. 216, 951–960.

Winter, H., Huber, S.C., 2000. Regulation of sucrose metabolism in higher plants: localization and regulation of activity of key enzymes. Crc. Cr. Rev. Plant Sci. 19, 31–67.

CHAPTER 7

Case Study: Trehalose

Chapter Outline

7.1 Introduction

Trehalose (α-D-glucopyranosyl-[1,1]-α-D-glucopyranoside) is a nonreducing disaccharide widely distributed in nature. It is a highly stable molecule due to its low free energy of hydrolysis ($\Delta G = -1.0$ kcal.mol^{-1}) compared with that of sucrose, and resistant to acids even at high temperatures (Birch et al., 1963; Fig. 7.1).

Trehalose is synthesized by many organisms, including bacteria, yeast, fungi, insects, invertebrates, green algae, and cyanobacteria; however, its synthesis in plants is still controversial (Elbein, 1974; Lee, 1980; Avigad, 1982; Klähn and Hagemann, 2011; Lunn et al., 2014). In general trehalose accumulates at high concentrations in organisms which have resistance and great capacity to tolerate stresses, reaching up to 30% of dry weight in fungal spore, rotifers, nematodes, adult tartidrades, crustacean embryonic cysts, and certain resurrection plants (Drennan et al., 1993). In insects, trehalose is the main hemolymph carbohydrate, fulfilling a role similar to that of glucose in the animal blood (Wyatt and Kalt, 1957). In other organisms, its general function is associated with that of a compatible solute that confers tolerance against different abiotic stresses. In *Sacharomyces cerevisiae*, there is evidence of a direct relationship between trehalose accumulation and the cell ability

Methods for Analysis of Carbohydrate Metabolism in Photosynthetic Organisms.
DOI: http://dx.doi.org/10.1016/B978-0-12-803396-8.00007-7

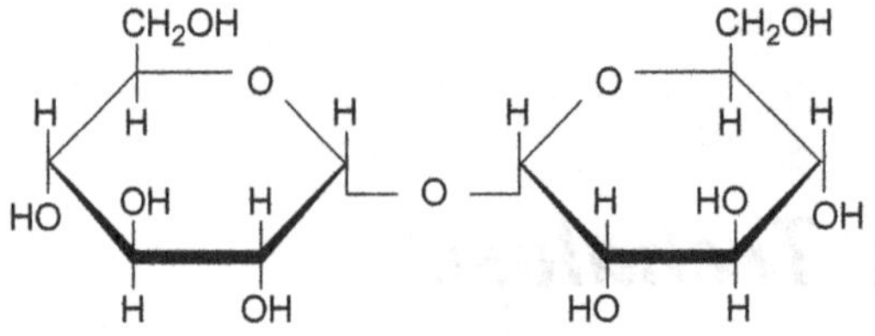

Trehalose

Figure 7.1
Trehalose structure (α-D-glucopyranosyl-[1,1]-α-D-glucopyranoside).

to cope with stresses (Gadd et al., 1987), such as freezing (Hino et al., 1990), high temperatures (Wiemken, 1990), alcohol (Mansure et al., 1994), and oxidative stress (Benaroudj et al., 2001; Pedreño et al., 2002). In *Escherichia coli*, trehalose has a similar role regarding salt and osmotic stress (Strøm and Kaasen, 1993). In addition to its function as a reserve carbohydrate, being a high-energy source that is freely available, it is a transport molecule and takes part in the cellular metabolism stabilizing proteins and membranes (Elbein et al., 2003).

In photosynthetic microorganisms, it is well documented that trehalose can accumulate in several species of cyanobacteria, algae, and free-living protists, often associated with adaptation to salt stress (Reed et al., 1984; Porchia et al., 1999; Klähn and Hagemann, 2011). On the contrary, easily detectable levels of trehalose are an exception in plants. Only in a small number of angiosperms resistant to desiccation (such as resurrection plants), trehalose represents a high percentage of leaf content in periods of extreme drought, protecting cellular structures from denaturation (Bartels and Hussain, 2011).

The capacity of vascular plants to produce trehalose has been a point of controversy because of the lack of its conclusive identification as an endogenous metabolite. In most plants, trehalose accumulates at a very low level (100–1000 times lower than that of sucrose) (Carillo et al., 2013). It was detected first in *Arabidopsis thaliana* using validamycin A, an inhibitor of trehalase (the enzyme that hydrolyzes trehalose) (Müller et al., 2001). Further studies extended this result to other plant species. Moreover, the presence of genes coding for trehalose-biosynthesis proteins (trehalose-phosphate synthase and trehalose-phosphate phosphatase) in *A. thaliana* and their molecular characterization opened new insights on trehalose functions in plants (Blázquez et al., 1998; Vogel et al., 1998; Leyman et al., 2001; Schluepmann et al., 2003). However, it is still necessary to fully demonstrate that the plethora of genes present in a plant genome catalyze the trehalose synthesis in vivo (Leyman et al., 2001). The fact that trehalose has been found in a few plants, and at very low levels, leads to the conclusion that the disaccharide does not have an important function as an osmoprotectant in relation to environmental stress protection. Even though the precise biological function of trehalose in plants is still unknown, it has been proposed that it may play a role in pathogenic and symbiotic plant–microbe interaction, and in responses to abiotic stresses (Carillo et al., 2013). Also, trehalose-6-phosphate, the intermediate in trehalose biosynthesis, is likely to have a role in signal transduction. A recent meta-analysis that revealed a parallel correlation between trehalose-6-phosphate level and sucrose, points to the existence of a bidirectional network, in which trehalose-6-phosphate is a signal of sucrose availability and acts to maintain sucrose concentrations within an appropriate range (Lunn et al., 2014). Mutants in trehalose-6-phosphate metabolism have highly pleiotropic phenotypes, showing defects in embryogenesis, leaf growth, flowering, inflorescence branching, and seed set. However, current models conflict with some experimental data, and do not completely explain the pleiotropic phenotypes exhibited by mutants in trehalose-6-phosphate metabolism.

7.2 Trehalose Biosynthesis and Degradation

At least five biosynthetic pathways have been described for trehalose in nature (Avonce et al., 2006). The most widely distributed pathway, initially demonstrated in yeast (Cabib and Leloir, 1958), is present in eubacteria, archaea, cyanobacteria, fungi, insects, and plants (Paul et al., 2008). It is a two-step pathway that involves two successive reactions catalyzed by trehalose-phosphate synthase (TPS, UDP-glucose, D-glucose-6-phosphate-1-glucosyltransferase, EC 2.4.1.15) and trehalose-phosphate phosphatase (TPP, trehalose-6-phosphate phosphohydrolase, EC 3.1.3.12):

$$\text{UDP-glucose} + \text{Glucose-6-phosphate} \xrightarrow{\text{TPS}} \text{UDP} + \text{Trehalose-6-phosphate}$$

$$\text{Trehalose-6-phosphate} \xrightarrow{\text{TPP}} \text{Trehalose} + \text{Orthophosphate (Pi)}$$

A second pathway involves the enzyme trehalose phosphorylase (TreP, α,α-trehalose phosphorylase EC 2.4.1.231), first described and studied by Marechal and Belocopitow (1972). TreP occurs in both prokaryotes including certain cyanobacterial strains and eukaryotes, including *Euglena gracilis*, a photosynthetic free-living protist (Fiol and Salerno, 2005). TreP catalyzes the reversible hydrolysis of trehalose in the presence of inorganic phosphate, generating glucose-1-phosphate and glucose.

$$\text{Trehalose} + \text{Pi} \xleftrightarrow{\text{TreP}} \text{Glucose} + \beta\text{-Glucose-1-phosphate}$$

A third synthetic pathway comprises the successive activities of malto-oligosyl trehalose synthase (TreY, EC 5.4.99.15), which catalyzes the conversion of maltopentaose into maltooligosyl trehalose by intramolecular transglycosylation, and maltooligosyltrehalose trehalohydrolase (TreZ) that hydrolyzes the maltooligosyl trehalose, releasing free trehalose. Putative TreY and TreZ encoding sequences were retrieved from a few cyanobacterial genomes (Klähn and Hagemann, 2011). In a fourth pathway, trehalose is produced by the isomerization of the 1α-α4 bond of maltose to a α1-α1 bond, forming trehalose by the action of trehalose synthase (TS). This enzyme was reported in bacteria and putative encoding sequences were found in a few cyanobacterial genomes (Klähn and Hagemann, 2011). A fifth biosynthetic pathway involves the reversible formation of trehalose from ADP-glucose and glucose catalyzed by a trehalose glycosyltransferase (TreT), which is found in bacteria and archaea (Paul et al., 2008) but in no photosynthetic organisms.

Trehalose degradation in photosynthetic organisms is carried out by the enzyme trehalase (TH, α,α trehalase, EC 3.2.1.28), found in insects, mammals, fungi, bacteria, and in many plant species (Müller et al., 2001). TH is highly specific and hydrolyzes trehalose to two glucose molecules. It is inhibited by validamycin A (Müller et al., 1995).

$$\text{Trehalose} + H_2O \xrightarrow{\text{TH}} 2\ \text{Glucose}$$

Trehalose-6-phosphate can also be degraded by the activity of trehalose-6-phosphate hydrolase (TPH, EC 3.2.1.93) or by the action of trehalose-6-phosphate phosphorylase (TrePP, α,α-trehalose-6-phosphate phosphorylase, EC 2.4.1.216). TPH was initially described in *E. coli* by Marechal (1984) and studied by Rimmele and Boos (1994). The enzyme catalyzes the following reaction:

$$\text{Trehalose-6-phosphate} + H_2O \xrightarrow{\text{TPH}} \text{Glucose} + \text{Glucose-6-phosphate}$$

TrePP, studied by Andersson et al. (2001) in *Lactococcus lactis*, catalyzes the reversible conversion of trehalose-6-phosphate into β-glucose-1-phosphate and glucose-6-phosphate. The equilibrium constant indicates that the reaction is displaced towards the formation of trehalose-6-phosphate. The enzyme catalyzes the following reaction:

$$\text{Trehalose-6-phosphate} + \text{Pi} \xleftrightarrow{\text{TrePP}} \beta\text{-Glucose-1-phosphate} + \text{Glucose-6-phosphate}$$

TPH and TrePP have not been reported in plants; however they could be found in cyanobacteria.

The most common enzymes related to trehalose metabolism in photosynthetic organisms are described below. Trehalose-6-phosphate hydrolase, trehalose phosphorylase, and trehalose-6-phosphate phosphorylase are included because they could be found in other photosynthetic microorganisms and are useful tools for the study of different aspects of trehalose metabolism.

Experimental Protocols

7.3 Trehalose Extraction

Trehalose can be extracted from trehalose-accumulating plants, algae, or cyanobacteria from freeze-dried or frozen material or cell pellets. Starting material (50–100 mg of fresh weight) is ground in a mortar to a fine powder under liquid nitrogen. The powder is suspended in 2 mL of alkaline water (brought to pH ~8 with ammonia solution) per gram of fresh weight. The suspension is heated at 100°C for 5 min under stirring with a glass rod and centrifuged at 12,000 × g for 5 min at 4°C. The supernatant is transferred to another tube. The pellet extraction is repeated twice more and the supernatants are combined and freeze-dried. Monosaccharides present in the sample can be destroyed by heating for 10 min in alkali 0.4 N. Trehalose is estimated by the anthrone method or after specific hydrolysis with addition of exogenous trehalase followed by the Somogyi–Nelson reaction (see Chapter 1: Determination of Carbohydrates Metabolism Molecules).

Comments

One serious limitation to the investigation of trehalose or trehalose-6-phosphate occurrence and function in most plants is that both carbohydrates are generally at very low concentrations. Successful extraction and determination methods have been described recently (Lunn et al., 2006; Carillo et al., 2013).

The extraction procedure for low amounts of trehalose or trehalose-6-phosphate (Lunn et al., 2006) starts with frozen material (10–20 mg of fresh weight) that is ground under liquid nitrogen. The extraction is made in a safe-lock microcentrifuge tube with 250 μL of chloroform:methanol (3:7, v/v) shaking and incubating at −20°C for 2 h. After addition of water and shaking at 4°C, the soluble sugars partition to the water–methanol phase.

Additionally, most of the trehalose assays are not sufficiently sensitive to detect the disaccharide (in a range level of pico to nanomol per gram of fresh weight) in plant tissues. One useful method employs glucose oxidase after hydrolysis with trehalase. Free glucose formed could be determined by fluorometric quantification (Carillo et al., 2013).

7.4 Enzyme Activity Assays

TPS activity can be measured by quantifying any of the reaction products (UDP or trehalose-6-phosphate). UDP formed can be enzymatically determined with pyruvate kinase in the presence of dinitrophenylhidrazyne or through a lactic dehydrogenase reaction. Trehalose-6-phosphate can be estimated after destroying the remaining glucose-6-phosphate with alkali. Then, trehalose-6-phosphate is determined by the anthrone reagent (Porchia et al., 1999). An alternative procedure (Van Vaeck et al., 2001) is hydrolyzing the trehalose-6-phospate formed with exogenous trehalose-phosphate hydrolase (TPH) that produces glucose and glucose-6-phosphate. Both products can be estimated by both spectrophotometric and fluorometric methods. In the last case, sensibility is 10-fold higher. TPP is assayed by measuring the inorganic phosphate released from trehalose-6-phospate.

TH activity is determined by measuring the produced glucose by either of the methods described above. TPH is assayed by measuring glucose or glucose-6-phosphate. The products of TreP reaction (glucose and glucose-1-phosphate) are also determined spectrophotometrically or fluorometrically, converting glucose-1-phosphate into glucose-6-phosphate by the action of phosphoglucomutase.

TrePP is assayed by incubating trehalose-6-phosphate with inorganic phosphate and measuring glucose-1-phosphate and glucose-6-phosphate formed as described in Chapter 1, Determination of Carbohydrates Metabolism Molecules. Additionally, TrePP activity in both reaction directions can be determined by HPAEC (high-performance anion-exchange chromatography) (Andersson et al., 2001).

The methods described above are used to assay activities of purified enzymes. To determine some of the activities in crude extracts, radioactive substrates are often employed. For example, UDP-[^{14}C]-glucose is used to assay TPS activity. The labeled product (trehalose-6-phosphate) is dephosphorylated with alkaline phosphatase (free of phosphodiesterase activity) and the obtained labeled trehalose can be separated from the substrate by an anionic exchange column (see Chapter 6: Case Study: Sucrose) to be quantified.

7.4.1 TPS Activity Assay

7.4.1.1 Measurement of trehalose-6-phosphate

Principle

The assay is similar to that used for sucrose-phosphate synthase (SPS) determination (see Chapter 6: Case Study: Sucrose). The method is based on the removal of the remaining substrate (glucose-6-phosphate) by heating with alkali (Leloir and Cardini, 1956), and trehalose-6-phosphate is estimated by the anthrone reagent.

Reagents

UDP-glucose	50 mM
Glucose-6-phosphate	100 mM
$MgCl_2$	100 mM
NaF	1 M
Hepes-NaOH buffer (pH 7.5)	1 M
NaOH	0.5 M
Anthrone reagent (see Chapter 1: Determination of Carbohydrates Metabolism Molecules)	

Procedure

In a 50-μL total volume, mix 0.25 μmol of UDP-glucose, 0.5 μmol of glucose-6-phosphate, 1 μmol of NaF, 5 μmol of Hepes-NaOH buffer (pH 7.5), and an aliquot of the enzyme preparation. Incubate 10–20 min at 30°C and stop the reaction by adding 200 μL of NaOH 0.5 M. Complete the blank tube with UDP-glucose. Stir up the tubes in a vortex and heat for 10 min at 100°C. After cooling, add 800 μL of the anthrone reagent and stir up again. Heat the samples for 15 min at 100°C. Finally, after cooling, read the absorbance at 620 nm. The difference with the incomplete tube gives the amount of trehalose formed. This method allows detecting up to 100 nmol of trehalose (200 nmol of glucose).

7.4.1.2 Measurement of UDP

Principle

The assay is similar to that used for SPS determination (see Chapter 6: Case Study: Sucrose). The method is based on the measurement of UDP formed by pyruvate kinase, which catalyzes the phosphorylation of nucleoside diphosphate to nucleoside triphosphate by phosphoenolpyruvate. This reaction leads to the formation of pyruvate which is determined by a colorimetric reaction (2,4-dinitrophenylhydrazine reacts with pyruvic acid forming 2,4-dinitrophenylhydrazone) (see Chapter 1: Determination of Carbohydrates Metabolism Molecules).

7.4.1.3 Measurement of glucose and glucose-6-phosphate

Principle

This method is based on the hydrolysis of trehalose-6-phosphate to glucose and glucose-6-phosphate after removing the ester used as substrate by heating with alkali (Van Vaeck et al., 2001). Hexoses can be determined spectrophotometrically or fluorometrically using the glucose-6-phosphate dehydrogenase/NADP system. If glucose is phosphorylated previously to the determination of glucose-6-phosphate, the sensitivity of the method can be duplicated.

Reagents

UDP-glucose	50 mM
Glucose-6-phosphate	100 mM
$MgCl_2$	100 mM
Hepes-NaOH buffer (pH 7.5)	1 M
NaOH	0.5 M
Hepes-NaOH buffer (pH 7.0)	100 mM
Phosphotrehalase of *Bacillus subtilis*	

Glucose and glucose-6-phosphate determination

ATP	100 mM
$MgCl_2$	200 mM
$NADP^+$	100 mM
Tris-HCl buffer (pH 8.0)	1 M
Hexokinase	
Glucose-6-phosphate dehydrogenase from baker's yeast (*S. cerevisiae*)	

Procedure

In a 50-μL total volume, mix 0.25 μmol of UDP-glucose, 0.5 μmol of glucose-6-phosphate, 1 μmol of NaF, 5 μmol of Hepes-NaOH buffer (pH 7.5), and an aliquot of the enzyme preparation. Omit UDP-glucose in the blank tube. Incubate at 30°C for 10–20 min. Stop the reaction by adding 200 μL of 0.5 M NaOH. Complete the blank tube with UDP-glucose. Stir the tubes in vortex and heat at 100°C for 10 min. After cooling, add 100 μL of 1 M Hepes-NaOH buffer (pH 7.5). Stir samples again and add 50 μL of phosphotrehalase (in 100 mM Hepes-NaOH buffer (pH 7.0)). Incubate the tubes at 37°C for 5 min. Stop the reaction by heating at 100°C for 1 min. Determine glucose and glucose-6-phosphate by the hexokinase and glucose-6-dehydrogenase system, adding 6 μL of 100 mM ATP, 6 μL of 200 mM $MgCl_2$, 6 μL of 100 mM $NADP^+$, 10 μL of 1 M Tris-HCl buffer (pH 8.0), 5 μL of hexokinase, 5 μL of glucose-6-phosphate dehydrogenase, and 12 μL of H_2O to each of the tubes used in this procedure. Incubate at 37°C for 15 min. Increase the volume to 1 mL with H_2O. Read absorbance at 340 nm. Determination of NADPH formed can be measured fluorometrically according to Jones et al. (1977) (see Chapter 1, Determination of Carbohydrates Metabolism Molecules).

Comments

The quantities indicated above are calculated for 0.1 μmol of glucose and glucose-6-phosphate in a total volume of 50 μL. It should be noted that the amount of ATP and $NADP^+$ used must be always in excess.

7.4.1.4 Activity assay in crude extracts

Principle

TPS activity can be determined in crude extracts using radioactive UDP-glucose. Labeled trehalose-6-phosphate is dephosphorylated by addition of alkaline phosphatase and radioactive trehalose is separated from the substrate by passage through anionic exchange resin.

Reagents

Glycine-NaOH buffer (pH 10)	1 M
UDP-[^{14}C]-glucose (160,000–240,000 cpm.μmol^{-1})	50 mM
Glucose-6-phosphate	100 mM
NaF	1 M
$MgCl_2$	200 mM
Hepes-NaOH buffer (pH 6.5)	1M
Anion exchange resin Dowex Cl^- (AG 1) (200–400 mesh)–X4	
Alkaline phosphatase (phosphodiesterase free)	

Procedure

In a 50-μL total volume, mix 0.5 μmol of UDP-[^{14}C]-glucose (specific activity 160,000–240,000 cpm.μmol^{-1}), 0.5 μmol of glucose-6-phosphate, 1 μmol of NaF, 5 μmol of 1 M Hepes-NaOH buffer (pH 6.5), and an aliquot of the enzyme preparation enzyme. The blank tube has the same composition except for glucose-6-phosphate. Incubate the tubes at 30°C for 10–20 min. Stop the reaction by adding 10 μL of 1 M glycine-NaOH buffer (pH 10) and heat at 100°C for 1 min. After cooling, complete blanks with glucose-6-phosphate. Add 5 μL of alkaline phosphatase (10 U · mg^{-1} protein, phosphodiesterase free). Incubate the tubes at 37°C for 20 min and stop the reaction by adding 0.2 mL of cool water (ice temperature). Pass the reaction mixture through a Bio-Rad AG1-X4 Cl^- (200–400 mesh) column of 0.6 × 2 cm. Elute labeled [^{14}C]-trehalose released from labeled [^{14}C]-trehalose-6-phosphate by alkaline phosphatase with 2 mL of water and collect the eluate in a scintillation vial. Add scintillation liquid and count radioactivity in a scintillation spectrometer.

Comments

The method described above also applies to both partially purified and pure enzymes where maximum sensitivity is required.

7.4.2 Trehalase Activity Assay

Principle

This method is based on the determination of the released glucose by the Somogyi–Nelson method or by measuring the oxidation of glucose with glucose oxidase (Friedman, 1966; Uhland et al., 2000).

Reagents

Trehalose	20 mM
Potasium phosphate buffer (pH 6.0)	500 mM

Procedure

In a 100-μL total volume, mix 0.2–0.5 μmol of trehalose, 5 μmoles of phosphate buffer (pH 6.0), and an aliquot of the enzyme preparation. Incubate at 30°C for 15 min. Stop the reaction by adding the Somogyi–Nelson reagent and determine glucose formed (see Chapter 1: Determination of Carbohydrates Metabolism Molecules).

7.4.3 Trehalose Phosphorylase Activity Assay

Principle

TreP activity can be assayed by measuring either trehalose synthesis or its phosphorolysis. In the first case, inorganic phosphate formed is quantified and in the second one, reducing power is measured. This method was described by Marechal and Belocopitow (1972) in order to determine the enzyme activity in *E. gracilis*.

Reagents

Synthesis	
β-glucose-1-phosphate	50 mM
Glucose	1 M
Imidazole-HCl buffer (pH 7.0)	500 mM
Phosphorolysis	
Trehalose	500 mM
Imidazole-HCl buffer (pH 7.0)	500 mM
Phosphate buffer (pH 7.0)	500 mM

Procedure

Synthesis

In a 50-μL total volume, mix 5 μL of 500 mM imidazole-HCl buffer (pH 7.0), 25 μL of β-glucose-1-phosphate, 5 μL of glucose, and an aliquot of the enzyme preparation. Incubate at 37°C for 30 min. Stop the reaction by addition of the Fiske–Subbarow reagent (or Chifflet reagents) and determine Pi released (see Chapter 1: Determination of Carbohydrates Metabolism Molecules).

Phosphorolysis

In a 50-μL total volume, mix 5 μL of 50 mM imidazole-HCl buffer (pH 7.0), 5 μL of 500 mM phosphate buffer (pH 7.0), 5 μL of 500 mM trehalose, and an aliquot of the enzyme preparation. Incubate at 37°C for 30 min. Stop the reaction by adding the Somogyi–Nelson reagent and determine glucose formed (see Chapter 1: Determination of Carbohydrates Metabolism Molecules).

7.4.4 Trehalose-6-Phosphate Hydrolase Activity Assay

Principle

The enzyme activity can be assayed by measuring glucose or glucose-6-phosphate. Glucose can be determined by its reducing power or by oxidation with the glucose oxidase/peroxidase system (Van Vaeck et al., 2001), while glucose-6-phosphate can be measured by the glucose-6-phosphate dehydrogenase/NADP system (Rimmele and Boos, 1994).

Reagents

Trehalose-6-phosphate	25 mM in 50 mM Bis-Tris buffer (pH 7.0)

Procedure

In a 50-μL total volume, mix 25 μL of 25 mM trehalose-6-phosphate in 50 mM Bis-Tris buffer (pH 7.0), and an aliquot or the enzyme preparation. Incubate at 37°C for 1–5 min. Reaction is stopped by heating 1 min at 100°C. Glucose released is determined by the glucose oxidase/peroxidase system.

7.4.5 Trehalose-6-Phosphate Phosphorylase Activity Assay

Principle

TrePP activity is determined by measuring glucose-6-phosphate produced by the glucose-6-phosphate dehydrogenase/NADP system (Andersson et al., 2001).

Reagents

Trehalose-6-phosphate	14 mM
Potassium phosphate buffer (pH 7.0)	200 mM
$NADP^+$	16 mM
Glucose-6-phosphate dehydrogenase	

Procedure

In a 150-μL total volume, mix 7.5 μL of 14 mM trehalose-6-phosphate, 7.5 μL of 16 mM $NADP^+$, 3.8 $U \cdot mL^{-1}$ glucose-6-phosphate dehydrogenase, and an aliquot of the enzyme preparation. Incubate at 37°C for 1–5 min. Follow the reaction by NADPH appearance at 340 nm (see Chapter 1: Determination of Carbohydrates Metabolism Molecules).

7.5 Stoichiometry

Principle

When studying the stoichiometry of a reaction, substrates' and products' concentration must be determined at a certain time after the reaction is started. The decrease in the amount of substrates should match the amount of products formed. In the reaction catalyzed by trehalose-6-phosphate synthase, it should be verified that the decrease of UDP-glucose and glucose-6-phosphate is equal (in μmol) to UDP and trehalose-6-phosphate formation. Particularly, UDP-glucose is determined by the UDP-glucose dehydrogenase reaction and glucose-6-phosphate by the reaction catalyzed by glucose-6-phosphate dehydrogenase (see Chapter 1: Determination of Carbohydrates Metabolism Molecules). Trehalose-6-phosphate and UDP formed are determined by the methods described above.

7.6 Isolation and Characterization of Reaction Products

The separation of trehalose and trehalose-6-phosphate could be achieved by anionic exchange chromatography in a resin in its borate form (as described for sucrose in Chapter 6: Case Study: Sucrose) or in a faster way, by high-performance liquid chromatography (HPLC) (in the order of minutes), which allows both analytic and preparative use.

The isolation of the products of trehalose-6-phosphate synthase and trehalose-6-phosphate phosphatase reactions (trehalose-6-phosphate and trehalose, respectively) involves their separation from the reaction substrates (UDP-glucose, glucose-6-phosphate, and UDP). The analytical separation could be achieved in 15 min using an anionic exchange CarboPac PA-100 (2×250 mm) column in a Dionex DX-300 system coupled to pulse amperometric detection. The column is equilibrated at least 10 min before using 0.1 N NaOH. The sample is eluted for 3 min with NaOH 0.1 N, followed by a 0–0.2 M sodium acetate gradient in 0.1 N NaOH for 3–8 min with a flow rate of 1 $mL \cdot min^{-1}$. In these conditions, retention time for trehalose and trehalose-6-phosphate are 2.8 and 7.9 min, respectively. Glucose-6-phosphate has a retention time of 10.4 min, while UDP-glucose and UDP have longer retention times, remaining on column. If glucose is present (product of the hydrolysis of glucose-6-phosphate, trehalose, or UDP-glucose), its retention time is 4.2 min, which does not interfere with the detection of the reaction products (De Virgilio et al., 1993). Separation could be achieved using a preparative column (22×250 mm) and collecting the eluent by a fraction collector. In order to recover the eluted substances (trehalose and trehalose-6-phosphate), a carbohydrate membrane desalter (CMD) is placed immediately after the amperometric detector. Its function is to desalt and lower the sample pH. More than 99% of sodium ions are removed by CMD. Even though the use of this membrane causes a reduction in the resolution (c.6%), it allows an acceptable purification. The maximum content of sodium ions is 0.35 M eluted at a speed of 1 $mL \cdot min^{-1}$. This membrane exchanges sodium by hydrogen ions. This process converts sodium hydroxide and sodium acetate in water and diluted acetic acid. In this way, the fractions containing the carbohydrates are free of salts. They can be freeze-dried and are ready to use in further studies.

HPLC columns can also be applied to the determination of enzymatic reactions; however, the number of reactions to simultaneously follow is limited.

Further Reading and References

Andersson, U., Levander, F., Rådström, P., 2001. Trehalose-6-phosphate phosphorylase is part of a novel metabolic pathway for trehalose utilization in *Lactococcus lactis*. J. Biol. Chem. 276, 42707–42713.

Avigad, G., 1982. Sucrose and other disaccharides. In: Loewus, F.A., Tanner, W. (Eds.), Encyclopedia of Plant Physiology, vol. 13A. Springer, Berlin, pp. 217–347. New Series.

Avonce, N., Mendoza-Vargas, A., Morett, E., Iturriaga, G., 2006. Insights on the evolution of trehalose biosynthesis. BMC Evol. Biol. 6, 109.

Bartels, D., Hussain, S.S., 2011. Resurrection plants: physiology and molecular biology. In: Lüttge, U., Beck, E., Bartels, D. (Eds.), Ecological Studies: Desiccation Tolerance in Plants. Springer, Heidelberg, pp. 339–364.

Benaroudj, N., Lee, D.H., Goldberg, A.L., 2001. Trehalose accumulation during cellular stress protects cells and cellular proteins from damage by oxygen radicals. J. Biol. Chem. 276, 24261–24267.

Birch, G.G., Wolfrom, M.L., Tyson, R.L., 1963. Advances of Carbohydrate Chemistry, vol. 18. Academic Press, New York.

Blázquez, M.A., Santos, E., Flores, C.L., Martínez-Zapater, J.M., Salinas, J., Gancedo, C., 1998. Isolation and molecular characterization of the *Arabidopsis TPS1* gene, encoding trehalose-6-phosphate synthase. Plant J. 13, 685–689.

Cabib, E., Leloir, L.F., 1958. The biosynthesis of trehalose phosphate. J. Biol. Chem. 231, 259–375.

Carillo, P., Feil, R., Gibon, Y., Satoh-Nagasawa, N., Jackson, D., Bläsing, O., et al., 2013. A fluorometric assay for trehalose in the picomole range. Plant Methods. 9, 21–35.

De Virgilio, C., Burckert, N., Jeno, P., Boller, T., Wiemken, A., 1993. Disruption of *TPS2*, the gene encoding the 100-kDa subunit of the trehalose-6-phosphate/phosphatase complex in *Saccharomyces cerevisiae*, causes accumulation of trehalose-6-phosphate and loss of trehalose-6-phosphate phosphatase activity. Eur. J. Biochem. 212, 315–323.

Drennan, P.M., Smith, M.T., Goldsworth, D., Van Staden, J., 1993. The occurrence of trehalose in the leaves of the desiccation tolerant angiosperm *Myrothamnus flabellifolius* Welw. J. Plant Phys. 142, 493–496.

Elbein, A.D., 1974. The metabolism of α-α-trehalose. Adv. Carbohyd. Chem. Biochem. 30, 227–256.

Elbein, A.D., Pan, Y.T., Pastuszak, I., Carroll, D., 2003. New insights on trehalose: a multifunctional molecule. Glycobiology. 13, 17r–27r.

Fiol, D.F., Salerno, G.L., 2005. Trehalose synthesis in *Euglena gracilis* (Euglenophyceae) occurs through an enzyme complex. J. Phycol. 41, 812–818.

Friedman, S., 1966. Trehalose-6-phosphatase from insects. In: Neufeld, E.F., Ginsburg, V. (Eds.), Methods Enzymology, vol. VIII. Academic Press, New York, pp. 372–374.

Gadd, G.M., Chalmers, K., Reed, R.H., 1987. The role of trehalose in dehydration resistance in *Saccharomyces cerevisiae*. FEMS Microbiol. Lett. 48, 249–254.

Hino, A., Mihara, K., Nakashima, Y., Takano, H., 1990. Trehalose levels and survival ratio of freeze-tolerant versus freeze sensitive yeasts. Appl. Environ. Microbiol. 56, 1386–1391.

Jones, M.G., Outlaw, W.H., Lowry, O.H., 1977. Enzymic assay of 10^{-7} to 10^{-14} moles of sucrose in plant tissues. Plant Physiol. 60, 379–383.

Klähn, S., Hagemann, M., 2011. Compatible solute biosynthesis in cyanobacteria. Environ. Microbiol. 13, 551–562.

Lee, C.K., 1980. Developments in Food Carbohydrate. Vol. 2. Disaccharides. Applied Science Publishers, London.

Leloir, L.F., Cardini, C.E., 1956. The biosynthesis of sucrose-phosphate. J. Biol. Chem. 214, 157–161.

Leyman, B., Van Dijck, P., Thevelein, J.M., 2001. An unexpected plethora of trehalose biosynthesis genes in *Arabidopsis thaliana*. Trends Plant Sci. 11, 510–513.

Lunn, J.E., Delorge, I., Figueroa, C.M., Van Dijck, P., Stitt, M., 2014. Small molecules: from structural diversity to signaling and regulatory roles. Trehalose metabolism in plants. Plant J. 79, 544–567.

Lunn, J.E., Feil, R., Hendriks, J.H.M., Gibon, Y., Morcuende, R., Osuna, D., et al., 2006. Sugar-induced increases in trehalose 6-phosphate are correlated with redox activation of ADPglucose pryophosphorylase and higher rates of starch synthesis in *Arabidopsis thaliana*. Biochem. J. 397, 139–148.

Mansure, J.J.C., Panek, A.D., Crowe, J.H., 1994. Trehalose inhibits ethanol effects on intact yeast cells and liposomes. Biochim. Biophys. Acta. 1191, 309–316.

Marechal, L.R., 1984. Transport and metabolism of trehalose in *Escherichia coli* and *Salmonella typhimurium*. Arch. Microbiol. 137, 70–73.

Marechal, L.R., Belocopitow, E., 1972. Metabolism of trehalose in *Euglena gracilis*. Partial purification and some properties of trehalose phosphorylase. J. Biol. Chem. 247, 3223–3228.

Müller, J., Boller, T., Wiemken, A., 1995. Trehalose and trehalase in plants: recent developments. Plant Sci. 112, 1–9.

Müller, J., Aeschbacher, R.A., Wingler, A., Boller, T., Wiemken, A., 2001. Trehalose and trehalase in Arabidopsis. Plant Physiol. 125, 1086–1093.

Paul, M.J., Primavesi, L.F., Jhurreea, D., Zhang, Y., 2008. Trehalose metabolism and signaling. Annual Rev. Plant Biol. 59, 417–441.

Pedreño, Y., Gimeno-Alcañiz, J.V., Matallaina, E., Arguelles, J.C., 2002. Response to oxidative stress caused by H_2O_2 in *Saccharomyces cerevisiae* mutants deficient in trehalase gene. Arch. Microbiol. 177, 494–499.

Porchia, A.C., Fiol, D.F., Salerno, G.L., 1999. Differential synthesis of sucrose and trehalose in *Euglena gracillis* cells during growth and salt stress. Plant Sci. 149, 43–49.

Reed, R., Richarson, D.L., War, S.L., Stewart, W.D., 1984. Carbohydrate accumulation in cyanobacteria. J. Gen. Microbiol. 130, 1–4.

Rimmele, M., Boos, W., 1994. Trehalose-6-phosphate hydrolase of *Escherichia coli*. J. Bacteriol. 176, 5654–5664.

Schluepmann, H., Pellny, T., van Dijken, A., Smeekens, S., Paul, M., 2003. Trehalose 6-phosphate is indispensable for carbohydrate utilization and growth in *Arabidopsis thaliana*. Proc. Nat. Acad. Sci. USA 100, 6849–6854.

Strøm, A.R., Kaasen, I., 1993. Trehalose metabolism in *Escherichia coli*: stress protection and stress regulation of gene expression. Mol. Microbiol. 8, 205–210.

Uhland, K., Mondigler, M., Spiess, C., Prinz, W., Ehrdmann, M., 2000. Determinants of translocation and folding of TreF, a trehalase of *Escherichia coli*. J. Biol. Chem. 275, 23439–23445.

Van Vaeck, C., Wera, S., Van Dijck, P., Thevelein, J.M., 2001. Analysis and modification of trehalose-6-phosphate levels in the yeast *Saccharomyces cerevisiae* with the use of *Bacillus substilis* phosphotrehalase. Biochem. J. 353, 157–162.

Vogel, G., Aeschbacher, R.A., Müller, J., Boller, T., Wiemken, A., 1998. Trehalose-6-phosphate phosphatases from *Arabidopsis thaliana*: identification by functional complementation of the yeast *tps2* mutant. Plant J. 13, 673–683.

Wiemken, A., 1990. Trehalose in yeast: stress protectant rather than reserve carbohydrate. Ant. Van Leeuwenhoek. J. Gen. Microbiol. 58, 209–217.

Wyatt, G.R., Kalt, G.F., 1957. The chemistry of insect hemolynph. Trehalose and other carbohydrates. J. Gen. Physiol. 40, 833–847.

CHAPTER 8

Case Study: Raffinose

Chapter Outline

8.1 Introduction

The trisaccharide raffinose (α-D-galactopyranosyl-(1 → 6)-α-D-glucopyranosyl-(1 → 2)-β-D-fructofuranoside) is the first member of a series of homologous oligosaccharides named "raffinose family oligosaccharide(s)" (RFO), which are α-1,6-galactosyl$_n$-sucrose (approximately $n < 7$) (Bachmann et al., 1994). The tetrasaccharide (degree of polymerization ($DP = 4$)) is known as stachyose, and the pentasaccharide and hexasaccharide, are named verbascose ($DP = 5$) and ajugose ($DP = 6$), respectively. The best-studied and most widespread polymers are the short-chain oligosaccharides (Fig. 8.1).

Raffinose ranks second to sucrose as the most common soluble sugar found in the plant kingdom. It is often found together with its higher homologs and galactinol. The relative proportions of these oligosaccharides vary between different plant species, but in general, raffinose and stachyose, and to a minor extent, verbascose, are the main sugars of this polymer series. In general, raffinose has been found mainly in leaves, stems, and storage organs (such as rhizomes, roots, and seeds) of a large number of legumes and other plants.

Methods for Analysis of Carbohydrate Metabolism in Photosynthetic Organisms.
DOI: http://dx.doi.org/10.1016/B978-0-12-803396-8.00008-9

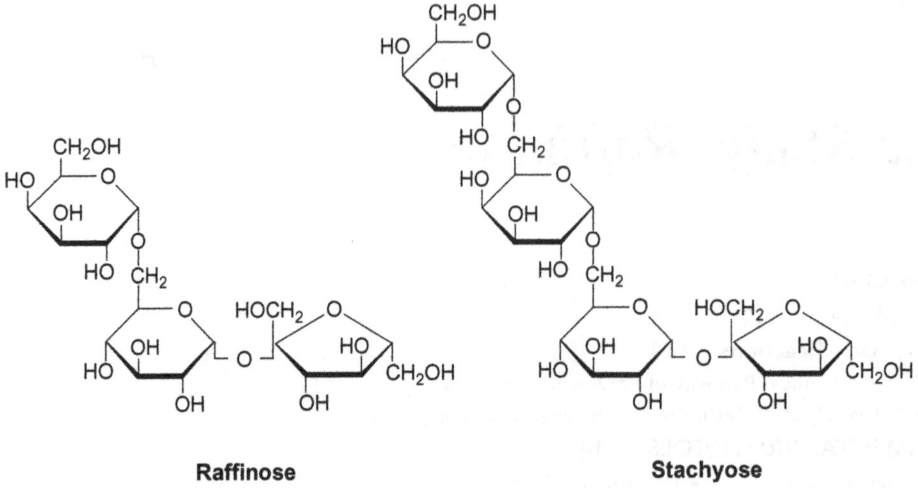

Raffinose

Stachyose

Figure 8.1
Raffinose (α-D-galactopyranosyl-(1→6)-α-D-glucopyranosyl-(1→2)-β-D-fructofuranoside) and stachyose (α-D-galactopyranosyl-(1→6)-α-D-galactopyranosyl-(1→6)-α-D-glucopyranosyl-(1→2)-β-D-fructofuranoside) structures.

It is mainly found in the plant cytosol (Dey, 1980), but it has also been detected in the chloroplast (Schneider and Keller, 2009; Knaupp et al., 2011). The distribution of RFO in plants has been the subject of discussion in many studies (Dey, 1980; Keller and Pharr, 1995; Sengupta et al., 2015). While in some plants (such as *Arabidopsis thaliana*) raffinose neither accumulates in large amounts nor is transported to other tissues or organs (Downie et al., 2003), in other plants (most monocotyledons), it is found in leaves and accumulates at high levels in seeds. The higher homologs of the raffinose series (stachyose and verbascose) accumulate in the seeds of dicotyledons (Peterbauer and Richter, 2001). On the other hand, raffinose and stachyose accumulate in many gymnosperms. Studies in *Picea excelsa* have shown seasonal variations of sucrose, raffinose, and stachyose in the pine needles. In summer, raffinose contents are negligible but in winter, cold acclimation and raffinose accumulation are strictly coupled (Kandler and Hopf, 1984).

Raffinose has been also found in unicellular green algae, such as *Chlorella vulgaris*, where it accumulates at chilling temperatures (Salerno and Pontis, 1989). So far, no record has been found of raffinose occurrence in cyanobacteria.

A wide range of functions have been attributed to this group of oligosaccharides in plants. In addition to serving as desiccation protectants in seeds, in some plant families they act as transport sugars in the phloem, as storage carbohydrates (as part of carbon partitioning strategies), and as signal transduction molecules (including biotic stress response) (Avigad and Dey, 1997; Sengupta et al., 2015). They also accumulate as compatible solutes in vegetative tissues as protective agents against various abiotic stresses such as freezing, drought, and high salinity (Kandler and Hopf, 1984; Keller and Pharr, 1995; Avigad and Dey, 1997; Peterbauer and Richter, 2001; Peters and Keller, 2009). The timing and accumulation of RFO are controlled for each abiotic stress through the differential expression of the isoforms responsible for the oligosaccharide biosynthesis (ElSayed et al., 2014).

8.2 Enzymatic Reactions

8.2.1 Biochemical Pathway of RFO Synthesis

The basic method of raffinose biosynthesis is through a trans-D-galactosylase catalyzed reaction. Initially, the galactose donor was suspected to be UDP-galactose. However, the widespread occurrence of galactinol (1-O-α-D-galactopyranosyl-L-*myo*-inositol) in various organs of raffinose-containing plants strongly suggested that it was involved in the biosynthesis of the trisaccharide. One of the main observations that pointed to galactinol as the galactosyl donor for raffinose synthesis was the fact that galactinol was only found in plants that contain RFO. In addition, the galactosyl residue of galactinol has all the kinetic characteristics of a galactose precursor for raffinose series synthesis (Tanner and Kandler, 1966; Senser and Kandler, 1967).

Galactinol synthase (UDP-galactose:*myo*-inositol galactosyltransferase, EC 2.4.1.123) catalyzes the first and key step for entering into the pathway of RFO biosynthesis, and plays a central role in inositol metabolism and in the regulation of carbon partitioning between sucrose and RFO (Saravitz et al., 1987). The second step involves raffinose synthase (galactinol:sucrose 6-galactosyl transferase, EC 2.4.1.82) that transfers the galactosyl moiety from galactinol to the C_6 of the glucose moiety of sucrose, yielding the trisaccharide raffinose ($DP = 3$). In a third step, the tetrasaccharide stachyose is produced by the action of stachyose synthase (galactinol-raffinose galactosyltransferase, EC 2.4.1.67) that transfers the galactosyl moiety from galactinol to the C_1 of the galactose moiety of raffinose. The galactosyl transfer reactions are reversible (Peterbauer and Richter, 2001). Stachyose synthase was shown in pea seeds to be a multifunctional enzyme that also catalyzes verbascose synthesis by galactosyl transfer from galactinol to stachyose ($DP = 4$) as well as by self-transfer of the terminal galactose residue from one stachyose molecule to another (Peterbauer et al., 2002). In fact, the biosynthesis of oligosaccharides of a higher degree of polymerization occurs via a galactinol-independent pathway, through the action of a galactan:galactan galactosyl transferase (Haab and Keller, 2002; Tapernoux-Lüthi et al., 2004). This is a unique enzyme that catalyzes the chain elongation of RFO by transferring a terminal α-galactosyl residue from one oligosaccharide molecule to another one. For example, verbascose ($DP = 5$) and ajugose ($DP = 6$) are produced when stachyose ($DP = 4$) is incubated with the enzyme galactan:galactan galactosyl transferase. In vitro, with verbascose as the substrate, this enzyme is able to elongate RFO at least up to $DP = 7$ from the intermediary product ajugose (Tapernoux-Lüthi et al., 2004).

The main reactions involved in the biosynthesis of the raffinose oligosaccharide family members are describe below.

$$\text{UDP-galactose} + myo\text{-inositol} \xrightarrow{\text{Galactinol synthase}} \text{Galactinol} + \text{UDP}$$

$$\text{Galactinol} + \text{Sucrose} \xleftrightarrow{\text{Raffinose synthase}} \text{Raffinose} + myo\text{-inositol}$$

$$\text{Galactinol} + \text{Raffinose} \xleftrightarrow{\text{Stachyose synthase}} \text{Stachyose} + myo\text{-inositol}$$

$$\text{Galactinol} + \text{Stachyose} \xleftrightarrow{\text{Stachyose synthase}} \text{Verbascose} + myo\text{-inositol}$$

$$\text{Stachyose} + \text{Stachyose} \xleftrightarrow{\text{Stachyose synthase}} \text{Verbascose} + \text{Raffinose}$$

$$\text{Stachyose} \xleftrightarrow{\text{Galactan:galactan galactosyl transferase}} \text{Verbascose (intermediate)} \leftrightarrow \text{Ajugose}$$

8.2.2 Hydrolysis of Raffinose Family Oligosaccharides

Alpha-galactosidases (α-galactoside galactohydrolase, EC 3.2.1.22) are the initial enzymes in RFO degradation pathway. They catalyze the hydrolysis of terminal, non-reducing α-D-galactose residues in α-D-galactosides. In addition, invertases (β-fructofuranosidase, EC 3.2.1.26) hydrolyze the fructose residue from the terminal sucrose (Dey, 1980; Avigad, 1982; Dey and del Campillo, 1984).

Alpha-D-galactosidases are classified as acidic or alkaline enzymes, in accordance with their optimum pH. Most of the plant α-D-galactosidases that have been studied are acidic enzymes (Keller and Pharr, 1995); however, α-D-galactosidases with optimum pH of 7.0–7.5 have been also described (Bachmann et al., 1994; Carmi et al., 2003).

Experimental Protocols

8.3 Extraction of Raffinose Polymers

8.3.1 Extraction from Leaf Material

Similarly to other water-soluble carbohydrate extraction (such as sucrose, trehalose, or fructans), raffinose and the other RFO polymers are extracted from plant material with ethanol (Sprenger et al., 1995). Leaf material (100–200 mg fresh weight) is successively extracted with three 0.6-mL portions, one portion each of 80% (v/v) ethanol, 50% (v/v) ethanol, and water at 80°C for 10 min each extraction. During each extraction, after heating, samples are cooled on ice for 1–2 min, and centrifuged at 15,000 × g for 5 min.

The supernatant extracts are pooled, and dried by vacuum centrifugation. The residue is suspended in 1 mL of water, centrifuged at 15,000 × g (5 min, 4°C), and the supernatant is desalted by passage through mixed-bed or Dowex 50-H and Amberlite IR4B-OH. Carbohydrates are analyzed by high-performance liquid chromatography-pulsed amperometric detector (HPLC-PAD) (Bachmann et al., 1994).

Comments

A similar procedure was described for soluble carbohydrate extraction from the resurrection plant *Xerophyta viscosa* (Peters et al., 2007) and from *A. thaliana* leaf material (Egert et al., 2013). For *Arabidopsis*, the biological material is flash-frozen in liquid nitrogen and macerated with a plastic pestle in an eppendorf tube. Carbohydrates are extracted at 80°C for 10 min twice by a three-step process: 1 mL per step of 80% (v/v), ethanol, 50% (v/v) ethanol, and distillated water, followed by centrifugation at 15,000 × g for 5 min at 4°C after each step.

8.3.2 Extraction from Algal Cells

Centrifuged algal cells (100 mg of fresh weight) are freeze-dried and extracted with three 0.6-mL portions of 80% (v/v) ethanol at 80°C for 10 min each extraction (Salerno and Pontis, 1989). Extracts are pooled and evaporated. The residue is suspended in water and centrifuged at 15,000 × g for 5 min at 4°C. The clear solution is desalted by passage through Dowex 50 (H^+) and IRA-4B (OH^-) columns. Carbohydrates are analyzed by HPLC-PAD, or alternatively, by descending chromatography on Whatman N°1 paper using phenol:water (4:1, v/v) or ethyl acetate:pyridine:water (12:5:4, v/v) as developing solvents (see Chapter 3: Protein and Carbohydrate Separation and Purification). Sugar position on the chromatogram is ascertained by comparison with standard positions located after developing with silver nitrate or bencidine reagents. After elution of the portion of paper, the sugar can be estimated directly or after the addition of invertase or α-galactosidase, according to methods previously described (see Chapter 1: Determination of Carbohydrates Metabolism Molecules).

8.3.3 Extraction from Seeds

RFO can be extracted from seeds (eg, from *Pisum sativum* seeds) (Ekvall et al., 2007). The material is ground to a powder. Oligosaccharides are extracted with 1.4 mL of ethanol 50% (v/v) per gram of material at room temperature for 30 min, stirring up in vortex every 10 min. After centrifugation at 3000 × g for 10 min, oligosacharides and proteins present in the supernatant are precipitated in ethanol 80% (v/v) at 80°C for 10 min. The solution is evaporated at 40°C to dryness and the residue is dissolved in distilled water. The solution is desalted by passage through mixed-bed or Dowex 50-H and Amberlite IR4B-OH, and the final eluate is freeze-dried. Raffinose polymers are separated by paper chromatography or thin layer chromatography (TLC) or analyzed by HPLC.

Comments

An additional method for RFO extraction using alkaline conditions can be used. The saccharide analysis is performed by HPLC with evaporative light scattering detection (ELSD). Stachyose, sucrose, and raffinose are simultaneously determined (Wang et al., 2013).

8.4 Galactinol Synthase Activity Assay

Principle

The activity of galactinol synthase is determined by incubating labeled UDP-galactose and *myo*-inositol (Bachmann et al., 1994). After retaining the unreacted substrate in an anion exchange resin, labeled galactinol is determined in a scintillation counter.

Reagents

UDP-[U-^{14}C]-galactose (1.85 kBq.pmol^{-1})	50 mM
myo-inositol	0.25 M
$MnCl_2$	10 mM
Dithiothreitol (DTT)	0.1 M
Hepes-NaOH buffer (pH 7.5)	0.5 M
Enzyme solution	
Dowex-1 resin (formate form) suspended in water (1 g.mL^{-1})	
Microfilter column and 0.45-μm Nylon-66 membrane	

Procedure

In a 100-μL total volume, mix 1 μL of 50 mM UDP-[U-^{14}C]-galactose (1.85 kBq.pmol^{-1}), 40 μL of 0.25 M *myo*-inositol, 10 μL of 10 mM $MnCl_2$, 10 μL of 0.1 M DTT, 10 μL of 0.5 M Hepes-NaOH buffer (pH 7.5), and an aliquot of the enzyme preparation. Incubate for 20 min at 30°C. Simultaneously incubate a reaction mixture lacking *myo*-inositol (to determine the radioactivity incorporated independently of the *myo*-inositol presence). Stop the reaction by adding 400 μL of ethanol. Retain unreacted UDP-[U-^{14}C]-galactose in an anion-exchange resin by adding to the reaction mixture 200 μL of Dowex-1 resin (formate form) resuspended in water in a proportion 1:1 (w/v) and shake for 30 min. Remove the resin by centrifugation through a microfilter column fitted with a 0.45-μm membrane filter of Nylon-66. Collect the filtrate directly into a scintillation vial, and count in a scintillation counter after the addition of 3 mL of scintillation cocktail.

8.5 Raffinose and Stachyose Synthase Activity Assays

Principle

Raffinose synthase and stachyose synthase activities are assayed by incubating galactinol with sucrose or raffinose, respectively, and the product formation is analyzed by HPLC (Bachmann et al., 1994).

Reagents

Galactinol	50 mM
Sucrose (for raffinose synthase)	0.4 M
Raffinose (for stachyose synthase)	0.4 M
Hepes-NaOH buffer (pH 7.5)	0.5 M

Procedure

In a 100-μL total volume, mix 10 μL of 50 mM galactinol, 10 μL of 0.4 M sucrose, 10 μL of 0.5 M Hepes-NaOH buffer (pH 7.5), and an aliquot of enzyme preparation. Incubate the tubes for 3 h at 30°C. Simultaneously, incubate a mixture lacking galactinol (reaction blank). Stop the reaction in a water bath at 100°C for 5 min. Cool the tubes and centrifuge at 14,000 × g for 5 min. Desalt supernatants before determination of raffinose by HPLC using a Sugar-Pak I column.

Similarly, stachyose synthase activity is assayed in a reaction mixture (100-μL total volume) containing 10 μL of 50 mM, 10 μL of 0.4 M, 10 μL of 0.5 M Hepes-NaOH buffer (pH 7.5), and an aliquot of enzyme preparation. The experimental procedure is identical to the described above for raffinose synthase.

8.6 Galactan-Galactan-Galactosyl Transferase Activity Assay

Principle

Galactan-galactan-galactosyl transferase activity assay is similar to that of stachyose synthase, except that galactinol is omitted in the reaction mixture. The incubation mixture contains as substrates the trisaccharide raffinose ($DP = 3$) or the tetrasaccharide stachyose ($DP = 4$). The analysis of the products is carried out after separation by HPLC (Haab and Keller, 2002). It should be noted that the protein homogenates should be prepared by acidic extraction (at pH 5.0).

Reagents

Raffinose (or stachyose)	100 mM
Citrate-phosphate buffer (McIlvaine buffer) pH 5.0	0.5 M
NaOH	0.5 M

Procedure

Desalt enzyme preparations in McIlvaine buffer pH 5.0, according to the procedure described in Chapter 2, Preparation of Protein Extracts. In a 100-μL total volume, mix 10 μL of 100 mM raffinose or stachyose, 20 μL of 0.5 M McIlvaine buffer (pH 5.0), and an aliquot of the desalted enzyme preparation. Incubate the mixture at 30°C for 15–60 min. Stop the reaction by addition of 5 μL of 0.5 M NaOH. Dilute samples with 25 μL of distilled water and analyze by HPLC using a CarboPak PA10 column (4 × 250 mm, Dionex, Sunnyvale, CA, United States). Formation of stachyose (from raffinose) or verbascose (from stachyose) is taken as a measure of the transferase activity.

8.7 Degradation of Raffinose Polymers by α-Galactosidase

Principle

The assay for α-galactosidase activity is based on the measurement of released galactose from RFO members, at acidic or alkaline pH. The acidic α-galactosidase activity is found in the vacuole and consequently, the assay is carried out at pH 5.0. The cytosolic α-galactosidase activity is assayed at pH 7.5. Free galactose produced is determined by the Somogyi–Nelson method (see Chapter 1: Determination of Carbohydrates Metabolism Molecules) or by using an enzyme-linked assay (Smart and Pharr, 1980).

Reagents

Raffinose or stachyose	100 mM
Citrate-phosphate buffer (McIlvaine buffer) (pH 5), for acidic α-galactosidases	1 M
Tris-HCl buffer (pH 7.5) for alkaline α-galactosidases	1 M

Procedure

In a 150-μL total volume, mix 50 μL of buffer (McIlvaine buffer, pH 5, for acidic α-galactosidases, or Tris-HCl buffer, pH 7.5, for alkaline α-galactosidases), 50 μL of 100 mM raffinose or stachyose, and 50 μL of desalted enzyme preparation. Incubate at 30°C for 20 min, and stop the reaction by adding Somogyi–Nelson's reagent to determine the released galactose (see Chapter 1: Determination of Carbohydrates Metabolism Molecules).

Comments

Assays for α-galactosidase activity can also be carried out routinely using 1 mM p-nitrophenyl-α-D-galactoside (instead of stachyose or raffinose) at acidic or alkaline pH.

Galactose may also be determined with the enzyme galactose oxidase (see Chapter 1: Determination of Carbohydrates Metabolism Molecules) or with β-galactose dehydrogenase (Bachmann et al., 1994).

8.8 Separation of Raffinose Polymers

RFO members can be separated by TLC or paper chromatography (Kotiguda et al., 2006) or by HPLC-PAD using a CarboPac PA1 column (Bachmann et al., 1994; Wang et al., 2013).

Further Reading and References

Avigad, G., 1982. Sucrose and other disaccharides. In: Loewus, F., Tanner, W. (Eds.), Plant Carbohydrates I. Springer, Berlin, Heidelberg, pp. 217–347.

Avigad, G., Dey, P.M., 1997. Carbohydrate metabolism: storage carbohydrates. In: Dey, P.M., Harborne, J.B. (Eds.), Plant Biochemistry. Academic Press, San Diego, CA, pp. 143–204.

Bachmann, M., Matile, P., Keller, F., 1994. Metabolism of the raffinose family oligosaccharides in leaves of *Ajuga reptans* L. (cold acclimation, translocation and sink to source transition: discovery of chain elongation enzyme). Plant Physiol. 105, 1335–1345.

Carmi, N., Zhang, G., Petreikov, M., Gao, Z., Eyal, Y., Granot, D., et al., 2003. Cloning and functional expression of alkaline α-galactosidase from melon fruit: similarity to plant SIP proteins uncovers a novel family of plant glycosyl hydrolases. Plant J. 33, 97–106.

Dey, P.M., 1980. Biosynthesis of planteose in *Sesamum indicum*. FEBS Lett. 114, 153–156.

Dey, P.M., del Campillo, E.D., 1984. Biochemistry of the multiple forms of glycosidases in plants. Ad. Enzymol. Rel. Areas Mol. Biol. 56, 141–249.

Downie, B., Gurusinghe, S., Dahal, P., Thacker, R.R., Snyder, J.C., Nonogaki, H., et al., 2003. Expression of a galactinol synthase gene in tomato seeds is up-regulated before maturation desiccation and again after imbibition whenever radicle protrusion is prevented. Plant Physiol. 131, 1347–1359.

Egert, A., Keller, F., Peters, S., 2013. Abiotic stress-induced accumulation of raffinose in Arabidopsis leaves is mediated by a single raffinose synthase (RS5, At5g40390). BMC Plant Biol. 13, 218.

Ekvall, J., Stegmark, R., Nyman, M., 2007. Optimization of extraction methods for determination of the raffinose family oligosaccharides in leguminous vine peas (*Pisum sativum*) and effects of blanching. J. Food Comp. Anal. 20, 13–18.

ElSayed, A.I., Rafudeen, M.S., Golldack, D., 2014. Physiological aspects of raffinose family oligosaccharides in plants: protection against abiotic stress. Plant Biol. 16, 1–8.

Haab, C.I., Keller, F., 2002. Purification and characterization of the raffinose oligosaccharide chain elongation enzyme, galactan: galactan galactosyltransferase (GGT), from *Ajuga reptans* leaves. Physiol. Plant. 114, 361–371.

Kandler, O., Hopf, H., 1984. Biosynthesis of oligosaccharides in vascular plants. In: Lewis, D.H. (Ed.), Storage Carbohydrates in Vascular Plants. Cambridge University Press, Cambridge, pp. 115–131.

Keller, F., Pharr, D.M., 1995. Metabolism of carbohydrates in sinks and sources: galactosyl-sucrose oligosaccharides. In: Zamski, E., Schaffer, A.A. (Eds.), Photoassimilate Distribution in Plants and Crops: Source-Sink Relationships. Marcel Dekker, New York, pp. 157–183.

Knaupp, M., Mishra, K.B., Nedbal, L., Heyer, A.G., 2011. Evidence for a role of raffinose in stabilizing photosystem II during freeze–thaw cycles. Planta. 234, 477–486.

Kotiguda, G., Peterbauer, T., Mulimani, V.H., 2006. Isolation and structural analysis of ajugose from *Vigna mungo* L. Carbohydr. Res. 341, 2156–2160.

Peterbauer, T., Richter, A., 2001. Biochemistry and physiology of raffinose family oligosaccharides and galactosyl cyclitols in seeds. Seed Sci. Res. 11, 185–197.

Peterbauer, T., Mucha, J., Mach, L., Richter, A.J., 2002. Chain elongation of raffinose in pea seeds. Isolation, characterization, and molecular cloning of a multifunctional enzyme catalyzing the synthesis of stachyose and verbascose. J. Biol. Chem. 277, 194–200.

Peters, S., Keller, F., 2009. Frost tolerance in excised leaves of the common bugle (*Ajuga reptans* L.) correlates positively with the concentrations of raffinose family oligosaccharides (RFOs). Plant Cell Environ. 32, 1099–1107.

Peters, S., Mundree, S.G., Thomson, J.A., Farrant, J.M., Keller, F., 2007. Protection mechanisms in the resurrection plant *Xerophyta viscosa* (Baker): both sucrose and raffinose family oligosaccharides (RFOs) accumulate in leaves in response to water deficit. J. Exp. Bot. 58, 1947–1956.

Salerno, G.L., Pontis, H.G., 1989. Raffinose synthesis in *Chlorella vulgaris* cultures after a cold shock. Plant Physiol. 89, 648–651.

Saravitz, D.M., Pharr, D.M., Carter, T.E., 1987. Galactinol synthase activity and soluble sugars in developing seeds of four soybean genotypes. Plant Physiol. 83, 185–189.

Schneider, T., Keller, F., 2009. Raffinose in chloroplasts is synthesized in cytosol and transported across the chloroplast envelope. Plant Cell Physiol. 50, 2174–2182.

Sengupta, S., Mukherjee, S., Basak, P., Majumder, A.L., 2015. Significance of galactinol and raffinose family oligosaccharide synthesis in plants. Front. Plant Sci. 6, 656.

Senser, M., Kandler, O., 1967. Galactinol, ein Galactosyldonor für die Biosynthese der Zucker der Raffinosefamilie in Blättern. Z Pflanzenphysiol. 57, 376–388.

Smart, E.L., Pharr, D.M., 1980. Characterization of α-galactosidase from cucumber leaves. Plant Physiol. 66, 731–734.

Sprenger, N., Bortlik, K., Brandt, A., Boller, T., Wiemken, A., 1995. Purification, cloning, and functional expression of sucrose:fructan 6-fructosyltransferase, a key enzyme of fructan synthesis in barley. Proc. Natl. Acad. Sci. USA 92, 11652–11656.

Tanner, W., Kandler, O., 1966. Biosynthesis of stachyose in *Phaseolus vulgaris*. Plant Physiol. 41, 1540–1542.

Tapernoux-Lüthi, E.M., Böhm, A., Keller, F., 2004. Cloning, functional expression, and characterization of the raffinose oligosaccharide chain elongation enzyme, galactan:galactan galactosyltransferase, from common bugle leaves. Plant Physiol. 134, 1377–1387.

Wang, Q.Q., Wang, P.F., Qui, M.J., Sun, P., Xu, D.W., Yang, S.H., et al., 2013. Optimization of extraction methods with alkali and determination of stachyose, sucrose, and raffinose in fresh rehmannia (*Rehmannia glutinosa* Libosh) using high-performance liquid chromatography with evaporative light scattering detection. J. Med. Plant. Res. 7, 2170–2176.

CHAPTER 9

Case Study: Fructans

Chapter Outline

9.1 Introduction

Fructans are polymers built up of fructose residues that constitute a series of homologous oligosaccharides carrying an α-D-glucosyl residue at the end of the chain, as in sucrose, via a β-(2→1) linkage. They are sucrose-derived oligosaccharides (Pontis, 1990).

The type of fructans is determined by the glycosidic linkage position of the fructose residues, which occurs at one of the two primary hydroxyls (β-(2→1) or β-(2→6) or both linkages). Respectively, they are classified into three major groups: (1) the inulin type, in which fructofuranosyl units are linked by β-(2→1) linkages. The first member of this series (degree of polymerization, $DP = 3$) is the trisaccharide 1-kestose (isokestose); (2) the levan (or phlein) type, with β-(2→6) fructosyl-fructose linkages. The trisaccharide corresponding to this series is called 6-kestotriose (early called 6-kestose or kestose), which is further elongated to polymers with β-(2→6)-linkages; and (3) the branched type (graminin) that contains both kinds of glycosidic linkages (β-(2→1) and β-(2→6)). There are two other

Methods for Analysis of Carbohydrate Metabolism in Photosynthetic Organisms.
DOI: http://dx.doi.org/10.1016/B978-0-12-803396-8.00009-0

minor groups of fructans with a non-terminal glucose, which are derivatives of the trisaccharide 6_G-kestotriose (originally called neokestose) (Fig. 9.1). These fructan series (called inulin neoseries (with 2→1 linkages) and levan neoseries (with 2→6 linkages)) are obtained by prolonging the chain at either of the two terminal fructoses of neokestose (Waterhouse and Chatterton, 1993; Livingston et al., 2009).

Regarding the chemical nature of fructans, they have a unique structural feature within the family of oligo- and polysaccharides: no bond of the fructose furanose ring is part of the macromolecular backbone (Marchessault et al., 1980). This means that from one furanose ring to another there is a $-CH_2O$ group that gives the fructan chain a mobility that is not found in any other carbohydrate chain.

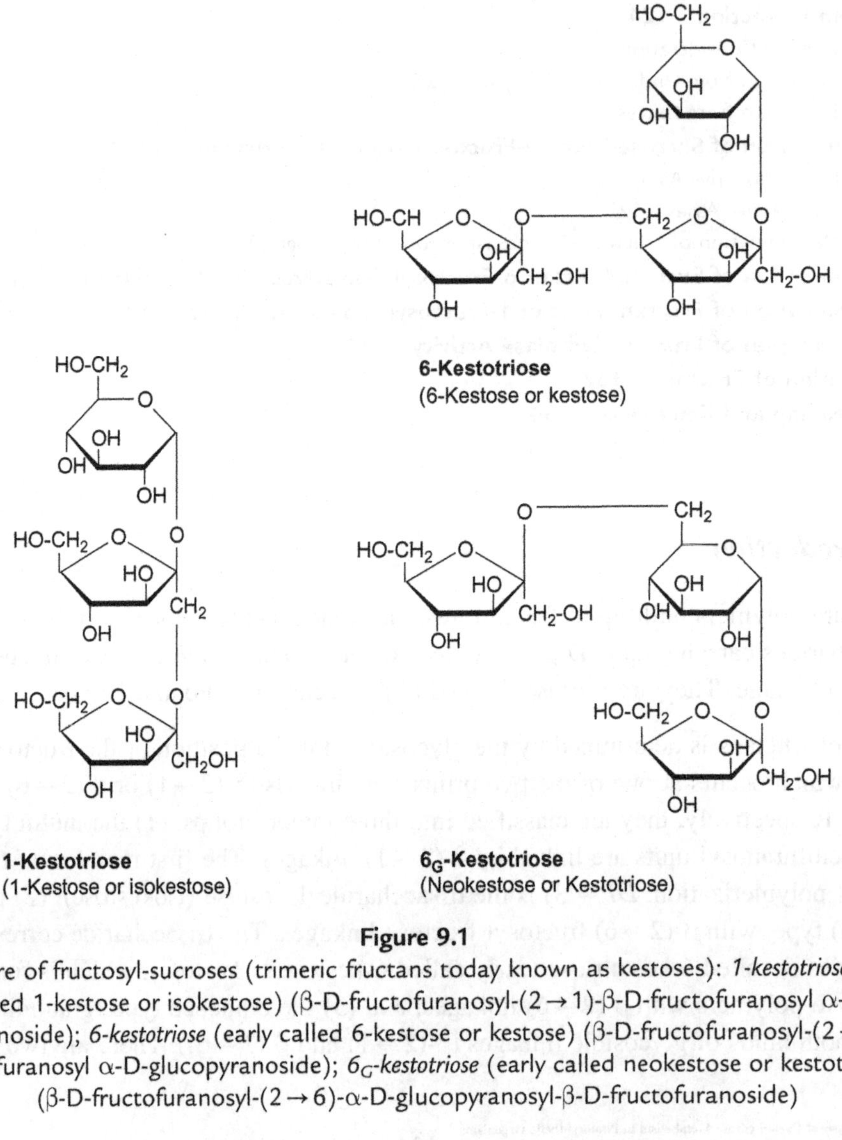

6-Kestotriose
(6-Kestose or kestose)

1-Kestotriose
(1-Kestose or isokestose)

6_G-Kestotriose
(Neokestose or Kestotriose)

Figure 9.1
Structure of fructosyl-sucroses (trimeric fructans today known as kestoses): *1-kestotriose* (early called 1-kestose or isokestose) (β-D-fructofuranosyl-(2→1)-β-D-fructofuranosyl α-D-glucopyranoside); *6-kestotriose* (early called 6-kestose or kestose) (β-D-fructofuranosyl-(2→6)-β-D-fructofuranosyl α-D-glucopyranoside); *6_G-kestotriose* (early called neokestose or kestotriose) (β-D-fructofuranosyl-(2→6)-α-D-glucopyranosyl-β-D-fructofuranoside)

Fructans are one of the most abundant water-soluble carbohydrates in the plant kingdom. They constitute the major storage carbohydrate in about 15% of flowering plant species (Hendry, 1993) and are present not only in monocotyledons and dicotyledons, but also in green algae and cyanobacteria (Pontis and Del Campillo, 1985; Hendry, 1993).

An outstanding characteristic of these oligosaccharides in plants is the fact that they can accumulate to very high concentrations. Values exceeding 50% tissue dry weight have been recorded for several members of the Compositae, Liliaceae, and Gramineae (Edelman and Jefford, 1968; Smith, 1973; Darbyshire and Henry, 1978; Wagner et al., 1983). In general, they are especially abundant in roots, bulbs, tubers, rhizomes, and some immature fruits (Meier and Reid, 1982) and their concentration in leaves is low. In some grasses, the lower sections of the stem are quite rich in fructans (Smith, 1973). Nevertheless, fructans concentration fluctuates considerably during the plant life cycle. At the subcellular level, these oligosaccharides appear to be located mainly in the vacuole (Frehner et al., 1984).

In monocotyledons (forage grasses such as *Dactylis glomerata*, *Phleum pratense*, *Poa secunda*, and cereals, such as wheat and barley), phlein-type and branched-group fructans are found in the aerial parts. In dicotyledons (such as *Helianthus tuberosus* and chicory), inulin-type fructans are mainly present in storage organs like tubers and rhizomes (Chatterton and Harrison, 1997; Van den Ende et al., 2003; Tamura et al., 2009). Recently, the synthesis of phlein-type fructans in a branch of dicotyledons, known as eudycotiledons, has been described. It should be mentioned that isokestose is generally present in both monocotyledons and dicotyledons (Wagner and Wiemken, 1987).

Among green algae, the Acetabularia family is rich in fructans, whose structure is similar to the one found in Compositae (inulin-type) (Smestad et al., 1972).

Fructans play a role in osmotic regulation, chilling, freezing, drought, and salinity tolerance, which is supported by the correlation between increasing fructans concentrations with stress tolerance (Pontis, 1989, 1990; Ritsema and Smeekens, 2003). Recently direct evidence confirmed the interaction between membranes and fructans acting as stabilizers either during freezing or drought (Livingston et al., 2009; Van den Ende, 2013).

9.2 Biosynthesis and Degradation of Fructans

Fructan biosynthesis takes place in two steps. The first one, general to all fructan-containing plants, is the biosynthesis of 1-kestose (1-kestotriose or β-2-fructosyl-sucrose) catalyzed by sucrose-sucrose-fructosyl transferase (SST) from two sucrose molecules. The second step depends on the fructan group. In Compositae, the enzyme that extends the fructan chain is fructan-fructan-fructosyl transferase (FFT), which adds fructosyl residues to the growing oligosaccharide by a β-(1→2) linkage. In Gramineae, the enzyme sucrose-fructan-fructosyl tranferase (SFT) catalyzes the synthesis of the first phlein, transferring a fructosyl residue from sucrose to the position 6 of the fructosyl residue of 1-kestose (1-kestotriose), thus starting the β-(2→6) linkages of the phlein fructan group. In certain plants there are fructosyl transferases that catalyze specific branching reactions.

The degradation of fructans occurs by the action of fructan hydrolases (exohydrolases), which are enzymes that shorten the hydrocarbon chain by a fructose unit. It must be pointed out that these enzymes attack neither sucrose nor fructosyl-sucrose.

The reactions involved in the biosynthesis and degradation of fructans are described below: sucrose-sucrose-fructosyl transferase (SST, EC 2.4.1.99); fructan-fructan 1-fructosyltransferase (1-FFT, EC 2.4.1.100); sucrose-fuctan 6-fructosyltransferase (6-SFT); and fructan hydrolase (FH, 3.2.1.80). In Compositae-family plants, "*n*" may be any member of the fructan series from 1 (trisaccharide) to approximately 35, and "*m*," for the acceptor molecule, is any member from 0 (sucrose) to approximately 35.

$$\text{Sucrose} + \text{Sucrose} \overset{\text{SST}}{\longleftrightarrow} \text{Glucose} + \text{1-Kestose (isokestose)}$$

$$\text{Sucrose-fructose}_n + \text{Sucrose-fructose}_m \overset{\text{FFT}}{\longleftrightarrow} \text{Sucrose-fructose}_{n-1} + \text{Sucrose-fructose}_{m+1}$$

$$\text{Sucrose} + \text{1-Kestose} \overset{\text{6-SFT}}{\longleftrightarrow} \text{Sucrose-fructose}_2\text{(bifurcose)} + \text{Glucose}$$

$$\text{Sucrose-fructose}_n \xrightarrow{\text{FH}} \text{Sucrose-fructose}_{n-1} + \text{Fructose}$$

Experimental Protocols

9.3 Fructan Extraction

9.3.1 General Considerations

To prevent alterations on the composition and the total amount of fructans, the biological material should be extracted immediately after collecting. Alternatively, the material can be plunged into liquid nitrogen and stored at −80°C, or freeze-dried until extraction. Drying the material at low temperatures in an oven is not an option, since enzymes can be still active and drying around 105°C causes changes in fructan contents (Laidlaw and Wylam, 1952; Raguse and Smith, 1965).

Fructans, together with other soluble sugars, are generally extracted with boiling water, but also with aqueous alcohol (ethanol or methanol) solution in a warm bath (eg, at 80°C or 90°C for 5 min), depending on the molecular weight of fructans. The use of alcoholic solvents shortens the extraction time and a more aqueous polar solvent benefits the extraction of monosaccharides, sucrose, and low molecular weight oligosaccharides. Additionally, the extraction with ethanol 80% (v/v) at 80°C serves both to inactivate enzymes and to extract sugars. The extraction of a low-molecular-weight fructan series with boiling alcohol could affect an arbitrary fraction of the polymers. In this case, an alcoholic extraction must be combined with a subsequent water extraction (Archbold, 1938).

9.3.2 Extraction from Underground Organs

The procedures described below with minor variations have been used to isolate fructans from Jerusalem artichoke tubers, dandelion, chicory, asparagus roots, and onion bulbs (Pontis, 1990).

Tubers, roots, bulbs, or rhizomes are washed, peeled, cut into small pieces, ground in a blender in ethanol 80% (v/v), and subsequently extracted three times by heating in a water bath. A small amount of calcium carbonate is usually added to the alcoholic solution to maintain a slightly alkaline pH during extraction. Care must be taken throughout the procedure in order to prevent fructan hydrolysis in acid pH. Supernatants are pooled, concentrated to near dryness *in vacuo* (removing the ethanol), and resuspended in water. The ethanol extract contains glucose, fructose, sucrose, and fructans of $DP < 7$. The residue obtained after the ethanol extraction is further extracted twice with boiling water. The two water extracts are combined and mixed with the aqueous solution obtained after dissolving the residue of the ethanol extraction. The resulting combined solution is passed through a mixed-bed ion exchanger column. If the solution needs clarification, centrifuge at 5000 × g for 10–20 min, and freeze-dry. The powder obtained contains glucose, fructose, sucrose, and fructans.

Comments

To extract fructans from small amounts of tissue (1–10 g), the tissue is dropped in 5 volumes of boiling alkaline water (pH ~8.0) or of hot (80°C) ethanol 80% (v/v).

The yield of fructans varies according to the stage of the plant development. From Jerusalem artichoke tubers collected at the beginning of autumn, a yield of up to 36% dry weight of total fructans may be obtained.

While qualitative extraction could not be considerably affected by the isolation method, the yield and ratios of the different saccharides (mono-, di-, oligo-, and polyfructooligosaccharides) could be affected. As an example, the effect of the extraction procedure on sugar content determination has been thoroughly described for onion bulbs (Davis et al., 2007).

9.3.3 Extraction from Leaves

After collection, leaves can be directly extracted or freeze-dried. In the case of fresh tissues, leaf segments of 0.5-cm long are immediately dropped into 5 volumes of boiling alkaline water (pH 8.0) and extracted for 5 min at 100°C. The extraction is repeated twice and every supernatant is collected in other tube. The combined extracts are freeze-dried. The residue is resuspended in one tenth volume of water and centrifuged at 5000 × g for 10 min for further analyses. In the case of freeze-dried leaves as a starting source material, samples are ground to powder under liquid nitrogen and extracted with boiling solvent as previously described.

Comments

When the equipment is available, fructans can be extracted in an Accelerated Solvent Extractor System from ground plant material (100 mg), as described for ryegrass (Abeynayake et al., 2015).

9.4 Determination of Sucrose-Sucrose-Fructosyl Transferase Activity

9.4.1 Nonradioactive Assay

Principle

This SST activity assay is based on the removal of unreacted sucrose molecules with a specific sucrase, which does not hydrolyze fructosyl-sucrose (1-kestose). The monosaccharides produced by the hydrolysis of sucrose and the glucose formed in the reaction catalyzed by SST are destroyed by heating with alkali. Fructosyl-sucrose produced is determined by estimation of fructose amounts (Puebla et al., 1999).

Reagents

Sucrose	1 M
Sodium maleate buffer (pH 6.5)	1 M
Acetate buffer (pH 5.2)	1 M
Sucrase	0.25 U per reaction
TBA-HCl reagent	

Note: dissolve the entire content of 1 vial of sucrase (Megazyme Internacional Ireland, Co. Wicklow, Ireland) in 22 mL of sodium maleate buffer 100 mM (pH 6.5). The solution is divided in aliquots of appropriate volume, which are stored at −20°C (stable for 2 years at this temperature).

Procedure

SST reaction mixture (50-μL total volume) contains 10 μL of 1 M sucrose, 5 μL of acetate buffer (pH 5.2), and an aliquot of the enzyme preparation. Omit sucrose in the blank tubes. Incubate at 30°C for an appropriate time (check up linearity up to 4 h). Stop the reaction by heating for 2 min at 80°C. Add 1–20 μL of each reaction mixture to tubes containing 0.25 U sucrose, completing up to a 0.4 mL final volume with 50 mM sodium maleate buffer (pH 6.5). Allow to proceed the hydrolysis of the unreacted sucrose for 1 h at 40°C. Cool sucrose treated samples at room temperature, and heat aliquots with 0.4 M NaOH in a final volume of 250 μL for 10 min at 100°C. Determine the amount of fructosyl-sucrose formed by the TBA-method (see Chapter 1: Determination of Carbohydrates Metabolism

Molecules), by adding 0.6 mL of TBA-HCl reagent and heating for 7 min at 100°C. After cooling, measure the absorbance at 432 nm against the reagent blank.

Comments

Alternatively, after incubation, the reaction is stopped by heating at 80°C for 5 min. Glucose released is assayed with the glucose oxidase peroxidase method, measuring both SST and invertase activity (Van den Ende and Van Laere, 1993).

9.4.2 Radioactive Assay

Principle

SST activity is determined by incubation with radioactive sucrose. After separating the products and sucrose by paper chromatography, the zone corresponding to the trisaccharide is cut and radioactivity is counted by scintillation spectrometry (Tognetti et al., 1988).

Reagents

[^{14}C]-sucrose (specific activity 6000 bq.μmol^{-1})	100 mM
Acetate buffer (pH 5.5)	1 M
Ethyl:acetate:pyridine:water (8:2:1, v/v/v)	

Procedure

SST enzyme activity is assayed in a reaction mixture (120-μL total volume) containing 100 μL of [^{14}C]-sucrose (specific activity 6000 bq.μmol^{-1}), 5 μL of acetate buffer (pH 5.5), and an aliquot of enzyme preparation. Omit the enzyme preparation aliquot in blank tubes. Incubate the mixture for 3 h at 30°C and stop the reaction by putting down aliquots of the reaction mixture on Whatman N°4 paper. Fructosyl-sucrose formed in the reaction is separated from unreacted sucrose by paper chromatography developed with ethyl:acetate: pyridine water (8:2:1, v/v/v). Chromatography is carried out at room temperature for approximately 12–14 h. The paper area corresponding to fructosyl-sucrose is cut and counted in a scintillation spectrometer.

Comments

To avoid the possible hydrolysis of fructosyl-sucrose or sucrose at pH 5.5, enzymatic reaction must not be stopped by heating at 100°C.

9.4.3 Determination of Fructosyl-Sucrose After Ion Chromatography Separation

Principle

The fructosyl-sucrose synthesized from sucrose by the action of SST can be analyzed and quantified by ion exchange chromatography (HPLC) using Dionex columns in an alkaline environment (Chatterton et al., 1989a). This method also allows the separation of the three fructosyl-sucrose isomers: isokestose, neokestose, and kestose. After incubation, the reaction mixture is loaded on a Dionex column, which separates glucose, sucrose, and fructosyl-sucroses (Fig. 9.2).

Reagents

Sucrose	1 M
Potassium citrate buffer (pH 5.5)	0.4 M
Bovine serum albumin (BSA)	
Saturated Ag_2SO_4 solution	
Ultracentrifuged enzyme preparation	
Dionex chromatograph and anion exchange carbohydrate column system	

Note: enzyme extract must be centrifuged at 12,000 × g for 20 min and the supernatants recentrifuged at 85,000 × g for 20 min.

Procedure

In a 200-μL total volume, mix 20 μL of 1 M sucrose, 0.2 mg BSA, 10 μL of 0.4 M potassium citrate buffer (pH 5.5), and an aliquot of enzyme preparation. Incubate the reaction mixture for 1–4 h at 30°C. Terminate the reaction by adding 200 μL of saturated Ag_2SO_4 solution. Centrifuge at 12,000 × g for 5 min, Separate the reaction components in a Dionex chromatograph equipped with an anion exchange carbohydrate column, and measure products with an inline pulsed amperometric detector.

9.5 Determination of Sucrose-Fructan-6-Fructosyl Transferase (6-SFT) Activity

Principle

Sucrose-fructan-6-fructosyl transferase is assayed by incubating isokestose and sucrose and determining bifurcose formation by ion exchange chromatography (Roth et al., 1997).

Reagents

MES-NaOH buffer (pH 5.8)	200 mM
Sucrose	500 mM
Isokestose	500 mM
Centrifuged enzyme extract from barley leaves	

Note: enzyme extract must be centrifuged at 12,000 × g for 20 min and the supernatants recentrifuged at 85,000 × g for 20 min.

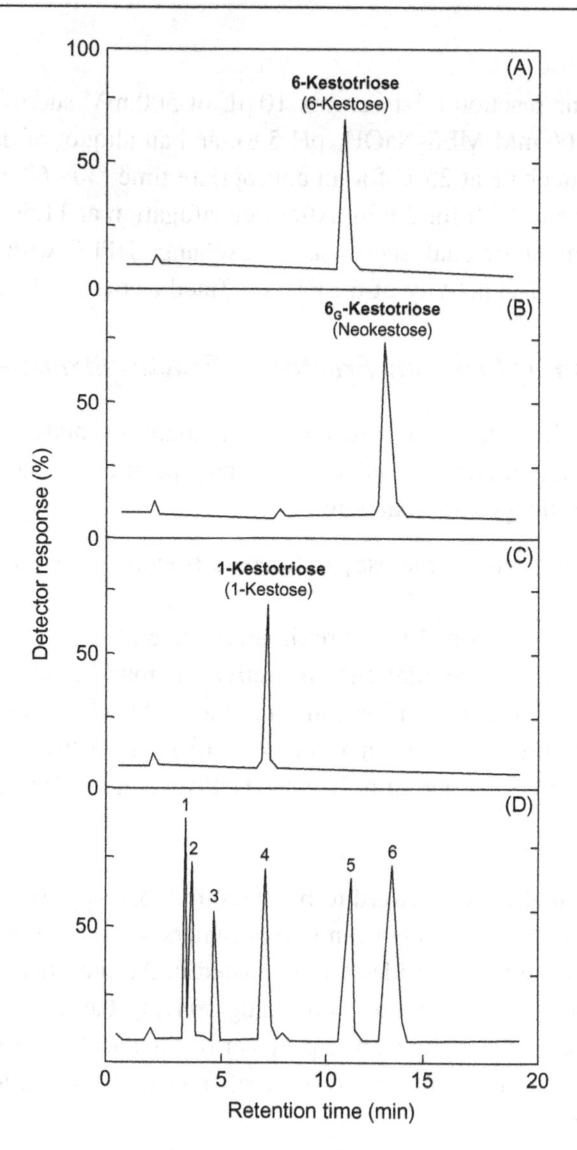

Figure 9.2

Fructosyl-sucroses purified by chromatography in AES-PAD. (A) 6-kestotriose (6-kestose) (23 $\mu g \cdot mL^{-1}$). (B) 6_G-kestotriose (neokestose) (34 $\mu g \cdot mL^{-1}$). (C) 1-kestotriose (1-kestose) (26 $\mu g \cdot mL^{-1}$). (D) Standard mixture consisting of 8 $\mu g \cdot mL^{-1}$ of glucose, fructose, and sucrose (peaks 1, 2, and 3, respectively) plus 1-kestotriose (peak 4), 6-kestotriose (peak 5), and 6_G-kestotriose (peak 6). Quantities are expressed as a percentage of detector response because the individual sugars have different deflections for a given quantity of sugar.

Procedure

In a 50-μL total volume reaction mixture, mix 10 μL of 500 mM sucrose, 10 μL of 500 mM isokestose, 10 μL of 100 mM MES-NaOH (pH 5.8), and an aliquot of the centrifuged enzyme preparation. Incubate at 25°C for an appropriate time (30–60 min) and stop the reaction in a boiling water bath for 2 min. After centrifugation at 11,500 × g for 3 min, aliquots of the supernatant are analyzed by anion-exchange HPLC with pulsed amperometric detection. The activity of 6-SFT is defined as the production of bifurcose.

9.6 Determination of Fructan: Fructan 1-Fructosyltransferase Activity

In general, 1-FFT activity enlarges a growing fructan chain by catalyzing the transfer of a single terminal β-D-fructofuranosyl residue to the same position of another fructan molecule, according to the general reaction:

$$\text{Sucrose-fructose}_n + \text{Sucrose-fructose}_m \leftrightarrow \text{Sucrose-fructose}_{n-1} + \text{Sucrose-fructose}_{m+1}$$

In *H. tuberosus*, "*n*" may be from 1 to approximately 35, and "*m*," may be varied from 0 (sucrose) to approximately 34. Similar enzyme activity is found in almost every member of the Compositae family (Scott et al., 1966; Shiomi et al., 1979; Van den Ende and Van Laere, 2007). Substrate preferences (donor and acceptor) vary between plant species, resulting in different patterns of inulin polymers (Hellwege et al., 1998).

Principle

1-FFT activity has been always awkward to be measured, because the enzyme catalyzes the redistribution of fructosyl residues between two polymers with different amounts of fructose. When the two oligosaccharides are incubated in the presence of 1-FFT, the change in the chain length can be observed after chromatographying the reaction mixture. The enzyme transfers a fructosyl residue from a polymer of higher *DP* to a fructan of lower *DP*. In practice, it is simpler to assay the transference of fructose from a fructan of $DP \geq 6$ to 1-kestose ($DP = 3$).

Reagents

Acetate buffer (pH 5.4)	1 M
1-Kestose	500 mM
Fructan ($DP = 7$)	500 mM
NaOH	0.5 M
Paper Whatman N°3	
Ethylacetate:pyridine:water (12:5:4, v-v:v)	
Naphtoresorcinol-HCl reagent (see Chapter 3: Protein and Carbohydrate Separation and Purification)	
Enzyme solution	

Procedure

In a 50-μL total volume, mix 10 μL of 1-kestose (5 μmol), 10 μL of fructan of $DP = 7$ (5 μmol), 5 μL of 1 M acetate buffer (pH 5.4), and an aliquot of the enzyme preparation. Incubate the reaction mixture for 3 to 5 h at 30°C. Stop the reaction by adding 10 μL of 0.5 M NaOH. Spot 25 μL of the incubation mixture on Whatman N°3 paper. Standards of 1-kestose ($DP = 3$) and fructans of $DP = 4$, 5, 6, and 7 are spotted at both sides of the incubation mixture spot. The chromatogram is run in ethyl:acetate:pyridine water for 18 h (Tognetti et al., 1988; Puebla et al., 1999). The paper is dried, sprayed with the naphtoresorcinol reagent, and heated for 3 min at 100°C. The activity of the enzyme is seen on the paper by the appearance of spots corresponding to fructans of $DP = 4$, 5, and 6.

Comments

To quantify the enzyme activity, the positions of the fructans present in the reaction mixture are ascertained by comparison with the position of the standards revealed with the naphtoresorcinol reagent. The fructans are eluted from the pieces of paper with water and quantified with the TBA method (see Chapter 1: Determination of Carbohydrates Metabolism Molecules).

It must be taken into consideration that the time of development of the chromatogram will depend on the chain length of the polymers used (Pontis, 1990).

Transferases similar to 1-FFT have been described in asparagus roots, onion bulbs, and ryegrass, but in these cases fructosyl residues are transferred to position 6 of the glucose or fructose moieties of sucrose (Ritsema and Smeekens, 2003; Abeynayake et al., 2015).

9.7 Determination of Fructan Hydrolase Activity

Principle

FH activity catalyzes the release of fructose residues from a fructan molecule according to the general reaction:

$$\text{Sucrose-fructose}_n + H_2O \xrightarrow{\text{FH}} \text{Sucrose-fructose}_{n-1} + \text{Fructose}$$

FH can be classified in two main groups of exohydrolases: (1) enzymes that hydrolyze β-(2→1) polymers (eg, FH isolated from barley leaves attacks β-(2→1) but also β-(2→6) fructosyl linkages) and (2) enzymes that hydrolyze β-(2→6) polymers (eg, FH isolated from orchard grass (*D. glomerata*) that shows high specificity for β-(2→6) linkages) (Yamamoto and Mino, 1985; Wagner and Wiemken, 1986). FH from both barley and *H. tuberosus* is likely located in the vacuole (Frehner et al., 1984; Wagner and Wiemken, 1986). The enzyme activity is measured by determining the fructose released in the reaction. In crude extracts, fructose estimation can be carried out by thin layer chromatography (TLC) which allows differentiating FH activity from SST and invertase products. On the other hand, to assay FH activity in purified enzyme preparations, the fructose produced is analyzed by its reducing power or by the use of auxiliary enzymes as it was described in general methods (see Chapter 1: Determination of Carbohydrates Metabolism Molecules). The assay of FH activity in barley leaves is described below.

Reagents

Citrate-phosphate buffer (McIlvaine buffer) (pH 5.2) 200 mM
Phlein (isolated from the stem base of *P. pratense*)

Procedure

FH activity is assayed in a mixture (200-μL total volume) containing 100 μL citrate buffer (pH 5.2), 2 mg of phlein, and an aliquot of the enzyme solution. Incubate the reaction mixture at 30°C for an appropriate time (30 min–2 h). Stop the reaction by boiling the mixture in a water bath for 3 min. Centrifuge at 12,000 × g (crude extract) and aliquots of this reaction are analyzed by TLC for the determination of the amount of reducing sugars (either by the Somogyi–Nelson method or by using auxiliary enzymes as described in Chapter 1: Determination of Carbohydrates Metabolism Molecules).

Comments

The procedure is essentially the same as described to assay FH enzymes from barley (Wagner and Wiemken, 1986) and orchard grass (Yamamoto and Mino, 1985). These enzymes do not hydrolyze either sucrose or kestose or isokestose.

9.8 Separation of Fructans

Fructans can be separated by their size or solubility. Resolution of fructan mixtures presents the difficulty that they exhibit similar composition and characteristics (two consecutive members of any fructan series differ in a single fructosyl residue). The first procedures used were based on the different solubility in alcohol solutions and descending paper chromatography (the preferred method several decades ago) is still used for fructan separation. Later, TLC began to be used to quantitatively assess fructans according to their molecular mass (Wagner and Wiemken, 1987). TLC is more advantageous than paper chromatography since it allows faster separations and resolves mixtures containing the three trisaccharides (kestose, isokestose, and neokestose). A clear separation of oligosaccharides up to nine fructosyl units can be obtained with both methods but they are not reliable when separating higher *DP* fructans (Shiomi, 1992; Benkeblia, 2013). Also, in the paper chromatograms or in the TLC plates, the position of each fructan is determined by staining with specific reagents for fructose. An additional advantage of paper chromatography is that it can show the real composition of higher *DP* polymers in terms of the total amount of fructose present in each fructan extract (Edelman and Dickerson, 1966).

Exclusion chromatography is another separation technique that can distinguish slight size differences between fructans. Initially it was developed by Pontis (1968) to separate fructans from Dahlia tubers by gel filtration on BioGel P-2. This method allows separating low molecular weight fructans up to $DP = 6-8$. Fructans are extracted from Dahlia tubers (collected at the end of summer) and from wheat leaves (harvested from 15-day-old seedlings kept at 4 °C for the last 7 days). Extracts are freeze-dried and resuspended in the minimum amount of water. An aliquot of the concentrated fructan solution is applied to a BioGel P-2 (200–400 mesh) column (220×1 cm), equilibrated with alkaline water (pH 8.0). This is an analytical column, but it can be turned into a preparative method (without losing resolution) by modifying the diameter of the column (but not the column height) and adjusting the other parameters in relation to the square of the diameter. Elution with alkaline water is performed at $10\ mL \cdot h^{-1}$ and 2-mL fractions are collected. Then, an aliquot of each fraction is analyzed by the TBA method in order to determine the presence of fructans. Fractions corresponding to each peak are mixed, freeze-dried, and stored at $-20°C$ (it is stable for several months) or at $-80°C$. This procedure is improved when using a BioGel P-2 column (less than 400 mesh) or by running the chromatography at 50°C. It has been used for the successful separation of fructans from onion (Darbyshire and Henry, 1978), *D. glomerata* (Pollock, 1982a), *Lolium temulentum* (Pollock, 1982b), wheat (Blacklow et al., 1984), and *H. tuberosus* (Cairns and Pollock, 1988).

BioGel chromatography is a method of choice for obtaining pure separated fructans for homologous series like those present in Dahlia or Jerusalem artichoke tubers. However, this method is not suitable for separating fructans of the same molecular weight belonging to different series (eg, fructans of β-(2$\rightarrow$1) and β-(2$\rightarrow$6) series found together as occurs in Gramineae). Moreover, the fractions corresponding to the trisaccharide peak may contain a mixture of kestoses and even raffinose (Fig. 9.3).

The resolution of fructan mixture separation has been enormously improved thanks to the use of strong alkaline conditions (pH $\geq$ 2) in anion-exchange columns. Most carbohydrates (pKa values $\sim$12–14) become anionic and can be separated on a basic anion exchanger in the hydroxide form. High performance anion-exchange chromatography (HPAEC) was used more than two decades ago to separate fructans using these columns (Chatterton et al., 1989a). Since then, this method, coupled with a pulse amperometric detector (PAD), has been used in the study of fructan composition of 180 selected Gramineae (Chatterton et al., 1989a,b), and for the analysis of oligosaccharides over a wide range of polymerization. The speed of the separation (15 min) and its resolution also make this method a powerful tool for assaying the enzymatic formation of fructosyl-sucrose. On the other hand, when it comes to studying the composition of every single fructan present in a plant tissue, the

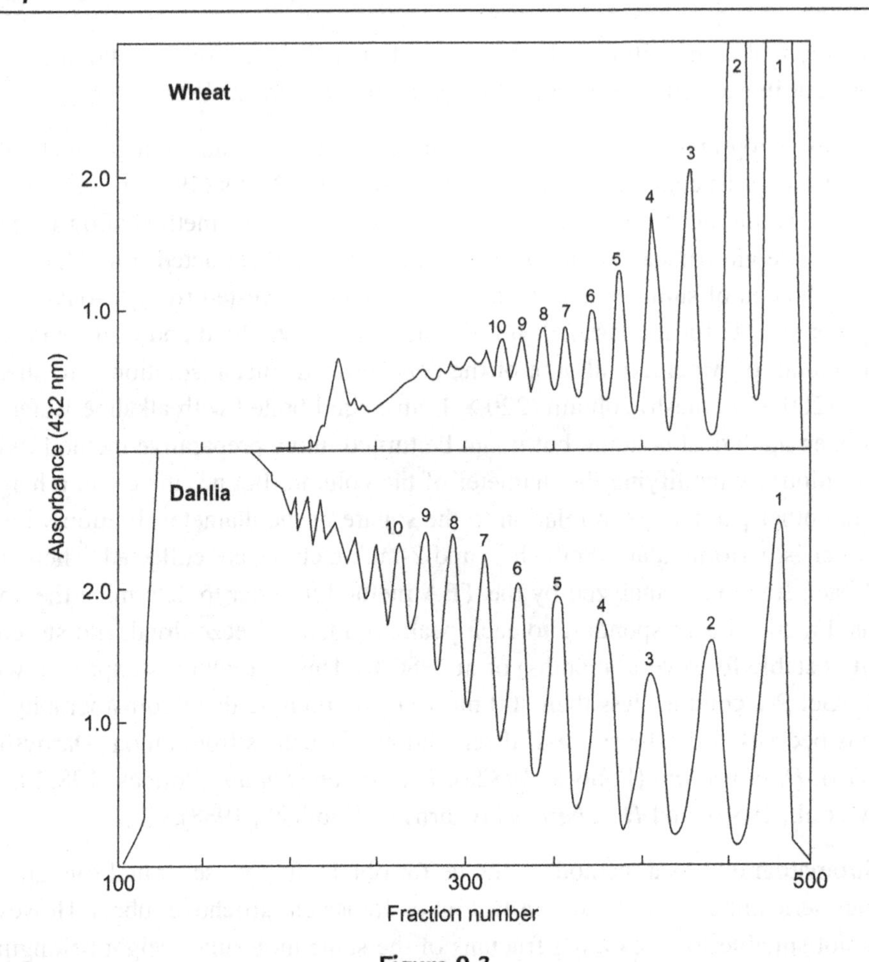

Figure 9.3
Separation of Wheat and Dahlia fructans by molecular exclusion chromatography. (1) fructose. (2) sucrose. (3–10) trisaccharide (fructosyl-sucrose) to the decasaccharide, respectively.

sensibility of the amperometric detector decreases with increasing *DP*. Consequently, the concentrations of fructans of higher *DP* are underestimated, and oligosaccharides of $DP > 20$ are barely detected (Vergauwen et al., 2003). Nevertheless, this technique has been improved by combining it with hydrolysis and an estimation of released short-chain fructans and fructose. A platform that combines HPAEC followed by the splitting of the effluent to the IPAD and to an on-line single quadrupole mass spectrometer (MS) allowed the detection of fructans up to $DP \sim 40$ (Bruggink et al., 2005).

Further Reading and References

Abeynayake, S.W., Etzerodt, T.P., Jonavičienė, K., Byrne, S., Asp, T., Boelt, B., 2015. Fructan metabolism and changes in fructan composition during cold acclimation in perennial ryegrass. Front. Plant Sci. 6, 1–13.

Archbold, H.K., 1938. Physiological studies in plant nutrition.VII. The role of fruns in the carbohydrate metabolism of the barley plant. Ann. Bot. 2, 183–187.
Benkeblia, N., 2013. Fructooligosaccharides and fructans analysis in plants and food crops. J. Chromatogr. A. 1313, 54–61.
Blacklow, W.M., Darbyshire, B., Pheloung, P., 1984. Fructans polymerized and depolymerized in the internodes of winter wheat as grain-filling progressed. Plant Sci. Lett. 36, 213–218.
Bruggink, C., Maurer, R., Herrmann, H., Cavalli, S., Hoefler, F., 2005. Analysis of carbohydrates by anion exchange chromatography and mass spectrometry. J. Chromatogr. A. 1085, 104–109.
Cairns, A.J., Pollock, C.J., 1988. Fructan biosynthesis in excised leaves of *Lolium temulentum* L. I. Chromatographic characterization of oligofructans and their labelling patterns following $^{14}CO_2$ feeding. New Phytol. 109, 399–405.
Chatterton, N.J., Harrison, P.A., 1997. Fructan oligomers in *Poa ampla*. New Phytol. 136, 3–10.
Chatterton, N.J., Harrison, P.A., Thornley, N.R., Bernett, J.H., 1989a. Purification and quantification of kestoses (fructosylsucroses) by gel permeation and anion exchange chromatography. Plant Physiol. Biochem. 27, 289–295.
Chatterton, N.J., Harrison, P.A., Thornley, N.R., Bernett, J.H., 1989b. Fructosyltransferase and invertase activities in leaf extracts of six temperate grasses grown in warm and cool temperatures. J. Plant Physiol. 135, 301–305.
Darbyshire, B., Henry, R.J., 1978. The distribution of fructans in onions. New Phytol. 87, 249–256.
Davis, F., Terry, L.A., Chope, G.A., Faul, C.F.J., 2007. Effect of extraction procedure on measured sugar concentrations in onion (*Allium cepa* L.) bulbs. J. Agric. Food Chem. 55, 4299–4306.
Edelman, J., Dickerson, A.G., 1966. The metabolism of fructose polymers in plants. Transfructosylation in tubers of *Helianthus tuberosus*. Biochem J. 98, 787–794.
Edelman, J., Jefford, T.G., 1968. The mechanism of fructosan metabolism in higer plants as exemplified in *Helianthus tuberosus*. New Phytol. 67, 517–531.
Frehner, M., Keller, F., Wiemken, A., 1984. Localization of fructan metabolism the vacuoles isolates from protoplasts of Jerusalem artichoke tubers (*Helianthus tuberosus* L.). J. Plant Physiol. 116, 197–208.
Hellwege, E.M., Raap, M., Gritscher, D., Wilmitzer, L., Heyer, A.G., 1998. Differences in chain length distribution of inulin from *Cynara scolymus* and *Helianthus tuberosus* are reflected in a transient plant expression system using the respective 1-FFT cDNAs. FEBS Lett. 427, 25–28.
Hendry, G.A.F., 1993. Evolutionary origins and natural functions of fructans-a climatological, biogeographic and mechanistic appraisal. New Phytol. 123, 3–14.
Laidlaw, R.A., Wylam, C.B., 1952. Analytical studies on the carbohydrates of grasses and clovers. II—the preparation of grass samples for analysis. J. Sci. Food Agric. 3, 494–497.
Livingston III, D.P., Hincha, D.K., Heyer, A.G., 2009. Fructan and its relationship to abiotic stress tolerance in plants. Cell. Mol. Life Sci. 66, 2007–2023.
Marchessault, R.H., Bleha, T., Deslandes, Y., Revol, J.F., 1980. Conformation and crystalline structure of (2→1)-β-d-fructofuranan (inulin). Can. J. Chem. 58, 2415–2422.
Meier, H., Reid, J.S.G., 1982. Reserve polysaccharides other than starch in higher plants. In: Pirson, A., Zimmermann, M.H. (Eds.), Encyclopedia of Plant Physiology, vol. 13A. Springer, Berlin and New York, pp. 418–471. New Series.
Pollock, C.J., 1982a. Patterns of turnover of fructans in leaves of *Dactylis glomerata* L. New Phytol. 90, 645–650.
Pollock, C.J., 1982b. Oligosaccharide intermediates of fructan synthesis in *Lolium temulentum*. Phytochemistry. 21, 2461–2465.
Pontis, H.G., 1968. Separation of fructosans by gel filtration. Anal. Biochem. 23, 331–333.
Pontis, H.G., 1989. Fructans and cold stress. J. Plan Physiol. 134, 148–150.
Pontis, H.G., 1990. Fructans. In: Dey, P.M. (Ed.), Methods in Plant Biochemistry, vol. 2. Academic Press Ltd., London, pp. 353–369.
Pontis, H.G., Del Campillo, E., 1985. Fructans: biochemistry of storage carbohydrates in green plants. In: Dey, P.M., Dixon, R.A. (Eds.), Biochemistry of Storage Carbohydrates in Green Plants. Academic Press, London, pp. 205–227.

Puebla, A., Battaglia, M.E., Salerno, G.L., Pontis, H.G., 1999. Sucrose-sucrose fructosyl transferase activity: a direct and rapid colorimetric procedure for the assay of plant extracts. Plant Physiol. Biochem. 37, 699–702.

Raguse, C.A., Smith, D., 1965. Forage analysis, carbohydrate content in alfalfa herbage as influenced by methods of drying. J. Agric. Food Chem. 13, 306–309.

Ritsema, T., Smeekens, S., 2003. Fructans: beneficial for plants and humans. Curr. Opin. Plant Biol. 6, 223–230.

Roth, A., Lüscher, M., Sprenger, N., Boller, T., Wiemken, A., 1997. Fructan and fructan-metabolizing enzymes in the growth zone of barley leaves. New Phytol. 136, 73–79.

Scott, R.W., Jefford, T.G., Edelman, J., 1966. Sucrose fructosyltransferase from higher plant tissues. Biochem. J. 100, 23–24.

Shiomi, N., 1992. Content of carbohydrate and activities of fructosyltransferase and invertase in asparagus roots during the fructo-oligosaccharide- and fructo- polysaccharide-accumulating season. New Phytol. 122, 421–432.

Shiomi, N., Yamada, J., Izawa, M., 1979. Synthesis of several fructo-oligosaccharides by asparagus fructosyltransferases. Agric. Biol. Chem. 43, 2233–2244.

Smestad, B., Percival, E., Bidwell, R.G.S., 1972. Metabolism of soluble carbohydrates in *Acetabularia mediterranea* cells. Can. J. Bot. 50, 1357–1361.

Smith, D., 1973. The non structural carbohydrates. In: Butler, G.W., Bailey, R.W. (Eds.), Chemistry and Biochemistry of Herbaceous. Academic Press, London, pp. 105–155.

Tamura, K., Kawakami, A., Sanada, Y., Tase, K., Komatsu, T., Yoshida, M., 2009. Cloning and functional analysis of a fructosyltransferase cDNA for synthesis of highly polymerized levans in timothy (*Phleum pretense* L.). J. Exp. Bot. 60, 893–905.

Tognetti, J.A., Calderon, P.L., Pontis, H.G., 1988. Fructan metabolism: reversal of cold acclimation. J. Plant Physiol. 134, 232–236.

Van den Ende, W., 2013. Multifunctional fructans and raffinose family oligosaccharides. Front. Plant Sci. 4. Available from: http://dx.doi.org/10.3389/fpls.2013.00247.

Van den Ende, W., Van Laere, 1993. Purification and properties of an invertase with sucrose: sucrose fructosyltransferase (SST) activity from the roots of *Cichorium intybus* L. New Phytol. 123, 31–37.

Van den Ende, W., Van Laere, A., 2007. Fructans in dicotyledonous plants: occurrence and metabolism. In: Norio, S., Noureddine, B., Shuichi, O. (Eds.), Recent Advances in Fructooligosaccharides Research. Research Signpost, pp. 1–14.

Van den Ende, W., Clerens, S., Vergauwen, R., Van Riet, L., Van Laere, Yoshida, M., et al., 2003. Fructan 1-exohydrolases. ß(2,1) trimmers during graminan biosynthesis in stems of wheat? Purification, characterization, mass mapping and cloning of two fructan 1-exohydrolase isoforms. Plant Physiol. 131, 621–631.

Vergauwen, R., Van Laere, A., Van den Ende, W., 2003. Properties of fructan:fructan 1-fructosyltransferases from chicory and globe thistle, two Asteracean plants storing greatly different types of inulin. Plant Physiol. 133, 391–401.

Wagner, W., Wiemken, A., 1986. Properties and subcellular localization of fructan hydrolase in the leaves of barley (*Hordeum vulgare* L. cv. Gerbel). J. Plant Physiol. 123, 429–439.

Wagner, W., Wiemken, A., 1987. Enzymology of fructan synthesis in grasses properties of sucrose-sucrose-fructosyltransferase in barley leaves (*Hordeum vulgare* L. cv Gerbel). Plant Physiol. 85, 706–710.

Wagner, W., Keller, F., Wiemken, A., 1983. Fructan metabolism in cereals: induction in leaves and compartmentation in protoplasts and vacuoles. Z. Pflanzenphysiol. 112, 359–372.

Waterhouse, A.L., Chatterton, N.J., 1993. Glossary of fructan terms. In: Suzuki, M., Chatterton, N.J. (Eds.), Science and Technology of Fructans. CRC Press, Boca Raton, pp. 1–7.

Yamamoto, S., Mino, Y., 1985. Partial purification and properties of phleinase induced in stem base of orchardgrass after defoliation. Plant Physiol. 78, 591–595.

CHAPTER 10

Case Study: Polysaccharides

Chapter Outline

10.1 Introduction

In general, polysaccharides are high-molecular weight molecules, composed of more than 10 repetitive units of carbohydrates (monosaccharides called monomers) linked by glycosidic bonds. Their structures vary from linear to highly branched molecules, which can have slight modifications within their monomer units. Their hydrolysis releases either the monomers or oligosaccharides. Polysaccharide properties depend on their structure and the nature of the repetitive unit, and, usually, differ from those of the monomers (ie, they may be amorphous or crystalline, more or less water soluble). They are classified into homopolysaccharides (or homoglycans, made by one type of monosaccharide) and heteropolysaccharides (or heteroglycans, constituted by more than one type of monomer).

Polysaccharides play an important role as storage molecules or as being part of structural components of plants, algae, and cyanobacteria. The main reserve polysaccharides in photosynthetic organisms are starch (in plants and algae) and glycogen (in cyanobacteria). In the plant kingdom, the main structural polymers are cellulose, mannans, galactomannans, and pectic polysaccharides from the primary cell wall. In macroscopic multicellular algae,

Methods for Analysis of Carbohydrate Metabolism in Photosynthetic Organisms.
DOI: http://dx.doi.org/10.1016/B978-0-12-803396-8.00010-7

three groups of polysaccharides predominate: alginic acid (in brown seaweeds), sulphated galactans (in red algae), and complex polysaccharides (in green seaweeds). In unicellular green algae, nonstarch polysaccharides (such as cell wall polysaccharides or exopolymers) are very diverse, and have been usually described from model strains or from particular species of interest in different fields of application.

Starch, glycogen, and cellulose (glucose homopolymers) are separately described as particular case studies (see Chapters 11, 12 and 13: Case Study: Starch, Case Study: Glycogen, and Case Study: Cellulose, respectively). The present chapter is devoted to other relevant polysaccharides in plants, algae, and cyanobacteria.

10.2 Plant Polysaccharides: Mannans

10.2.1 Structure and Classification

Plants and several algae produce a wide range of polysaccharides containing D-mannose. These are polymers composed by (1→4)-β-linked D-mannosyl units that are mainly involved in storage functions and in structural roles (Meier and Reid, 1982; Painter, 1983; Stephen, 1983). In general, they are described as mannose linear chains; however, they can contain small amounts (less than 5%) of D-glucosyl and D-galactosyl units. From a chemical point of view, mannan polysaccharides are divided in four main categories: mannans, glucomannans, galactomannans, and galactoglucomannans (Matheson, 1990). However, this classification is not strict with regard to the sugar content. Simplified formulas of these polysaccharides are depicted in Fig. 10.1.

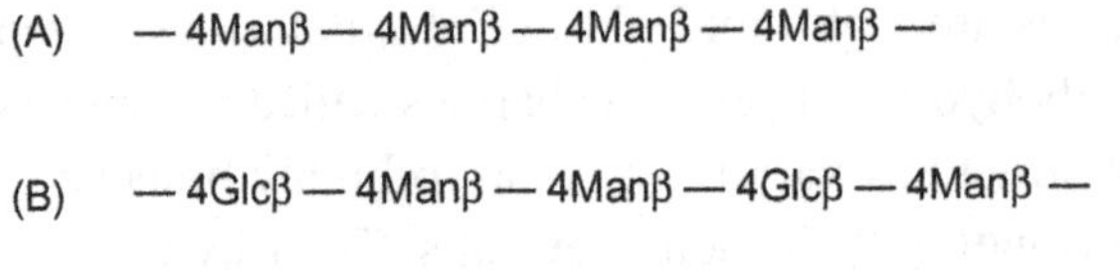

(A) —4Manβ—4Manβ—4Manβ—4Manβ—

(B) —4Glcβ—4Manβ—4Manβ—4Glcβ—4Manβ—

(C) —4Manβ—4Manβ(6←Galα)—4Manβ—4Manβ(6←Galα)—4Manβ—

(D) —4Manβ—4Glcβ—4Manβ(6←Galα)—4Manβ—4Glcβ—

Figure 10.1
Chemical structure of mannan polysaccharides. (A) (1→4)-β-mannan; (B) (1→4)-β-glucomannan; (C) (1→6)-α-galacto-(1→4)-β-mannan; (D) (1→6)-α-galacto-(1→4)-β-glucomannan. *Man*, mannose; *Glc*, glucose, and *Gal*, galactose residues.

Due to their structures, mannans are highly insoluble compounds and partial extraction is usually carried out either in alkaline conditions or with cuprammonium solution. In general, extracted mannans reach polymerization-degree values between 10 and 80 units.

10.2.2 Occurrence

Mannans are widespread among plants and are also present in many algal species. They are part of the plant cell wall playing a structural function as hemicelluloses that bind cellulose. Mannans are also structural compounds as crystalline fibrils in many algae that lack cellulose in their cell walls (Frei and Preston, 1968; Mackie and Preston, 1968). Also they serve as nonstarch carbohydrate reserves in endosperm walls, Orchidaceae pseudo bulbs and vacuoles (either in seed or vegetative tissue vacuoles) (Meier and Reid, 1982; Brennan et al., 1996). Additional potential important functions have been proposed for these polysaccharides (Liepman et al., 2007).

As mentioned above, structural polysaccharides as glucomannans are found as part of the cell wall, mainly being part of the hemicelluloses. In the wood of gymnosperms, they are the most abundant cross-linking glycans (up to 10%) (Maeda et al., 2000), while in angyosperms has been determined up to 3–5%. They have also been detected in a large number of monocotyledon seeds and in the endosperm of some Liliaceae, Iridaceae, Amaryllidaceae, and Agavaceae. These polymers are also found in tubers, bulbs, roots, and leaves of Liliaceae, Amaryllidaceae, Dioscoraceae, Orchidaceae, and Araceae. Among these, orchid and *Amorphophallus konjac* tubers are the best studied. The amount of glucomannans extracted from plant tissues may vary from a low percentage up to more than 30% depending on the source and nature of the tissue (Dey, 1978; Matheson, 1990).

Galactomannans are usually present in the endosperm of some legume seeds as extracellular deposits. They have been also reported in species of Annonaceae, Convolvulaceae, Ebenaceae, and Palmae, where they may be present from traces (as in soybeans) up to nearly 40% as in carob pods (Matheson, 1990).

In the case of galactoglucomannans, they form the hemicellulosic fraction of wood and in some gymnosperms they have been found in percentages up to 6%. Galactoglucomannans have been extracted from asparagus and legume seeds such as *Cersis siliquastrum* where they represent up to 10% of seeds' dry weight (Mills and Timell, 1963).

10.2.3 Biosynthesis and Degradation

Mannan biosynthesis is initiated by the transfer of a mannosyl, glucosyl, or galactosyl group from GDP-mannose, UDP-glucose, or UDP-galactose, respectively. The metabolic pathway leading to GDP-mannose (the mannosyl donor) synthesis starts with the isomerization of fructose-6-phosphate to mannose-6-phosphate (catalyzed by the enzyme mannose-6-phosphate isomerase, EC 5.3.1.8) followed by the conversion of mannose-6-phosphate into mannose-1-phosphate (catalyzed by the enzyme phosphomanno-mutase, EC 5.4.2.8). Then, mannose-1-phosphate reacts with GTP in a reaction catalyzed by the enzyme GDP-mannose pyrophosphorylase (EC 2.7.7.13) producing GDP-mannose and pyrophosphate. On the contrary, UDP-galactose formation is due to the action of UDP-glucose-4-epimerase (see Chapter 15: Case Study: Nucleotide Sugars).

The mannosyl-residue transfer from GDP-mannose is catalyzed by the enzyme mannan synthase (mannan 4 β-mannosyl transferase, EC 2.4.1.32). Two different enzymes have been identified: β-1,4-(gluco) mannan synthase that polymerizes the polysaccharide backbone, and α-1,6 galactomannan galactosyltransferase that adds galactose residue side chains to the backbone (Edwards et al., 1999). The degradation of mannose homopolymers is carried out by endo-hydrolysis due to a β-mannanase activity, while additional glucosidases and galactosidases act in the degradation of heteropolymers.

10.3 Algal Polysaccharides

In contrast to plants, where polysaccharides are mainly in the cell walls to give mechanical strength, in macroalgae the most abundant polysaccharides have the properties of gels or mucilages and those similar to cellulose are in lesser amounts. These types of polymers are also found in animals but not in land plants. Multicellular algae have one or more polysaccharides, in which some hydroxyl groups are esterified with sulfate groups. Their structure and composition depend on the group of algae.

Alginic acid, a linear polymer of (1→4)-linked mannuronic acid and guluronic acid (combined mainly with calcium) together with complex fucose-containing polymers and other sugar residues, and sulfate esters, occur in the cell walls and intercellular matrix of brown seaweeds. The saccharides are present in blocks of each monomer and the length and proportions of mannuronic and glucoronic residues of the blocks varies widely from one species to another, which results in alginates with different properties.

Sulfated galactans (agars and carrageenans) are the major polysaccharides of red algae (Rhodophyceae), and consist of linear chains of alternating sulfated (1→3)-β-galactose and (1→4)-α-galactose, 3,6-anhydrogalactose units and other polysaccharides. The third main group of algae, green seaweeds (Chlorophyceae), contains complex, usually sulfated polysaccharides comprising a number of different monosaccharides. Additionally, red and green algae synthesize starch-type materials, while brown algae synthesize laminarin, a β-(1,3)-glucan. The mentioned polysaccharides constitute a common feature among all the algae that synthesize glycans as storage polymers.

The extraction of the different polysaccharides specific to each type of macroalgae requires complex procedures which can be found in the following references: Jabbar-Mian and

Percival (1973), Carlberg and Percival (1977), Percival (1978), Craigie and Leigh (1978), and Percival et al. (1983). In most cases, after isolation and purification, these complex polysaccharides can be quantified by general methods (eg, the Dubois method described in Chapter 1: Determination of Carbohydrates Metabolism Molecules), and the components can be determined after the action of different hydrolytic enzymes (Percival and McDowell, 1981). When polysaccharides contain sulfate, particular procedures are required for their determination, which include precipitation with cetyl pyridinium chloride (Scott, 1960), precipitation with 4-chloro-4-amino-diphenyl (Jones and Letham, 1954), or as a barium sulfate precipitate (Dodgson and Prince, 1962).

The polysaccharides found in the diverse group of unicellular green algae (microalgae), as cell-wall components or extracellular polymers, are heteroglycans with different compositions. Polysaccharides from several *Chlamydomonas* and *Chlorella* strains have been well studied. For example, extracellular polysaccharides released by *Chlamydomonas* strains contain mainly arabinose, glucose, and galactose and are highly branched polymers composed of (1, 3, 4)-linked glucose and terminal arabinose and galactose units. The chemical composition of these polymers varies during growth. In the stationary phase of growth, glucose and glucuronic acid are the main components. The major fraction of the extracellular polysaccharides can be then described as a linear polymer composed of $(1 \rightarrow 4)$-linked glucose and $(1 \rightarrow 4)$-linked glucuronic acid. In contrast, extracellular polysaccharides from other *Chlamydomonas* strains (*Chlamydomonas corrosa*) are independent of the growth status and contain arabinose and galactose as major sugars (Allard and Tazi, 1993).

A detailed procedure for isolating cell walls and exocellular polymers of two *Chlamydomonas* strains (*Chlamydomonas mexicana* and *Chlamydomonas sajao*) has been reported by Barclay and Lewin (1985). Different attempts to extract and identify polysaccharides (other than starch), as well as two methods for the recovery of total cell-wall polysaccharides from *Chlorella* strains have been thoroughly described by Sui et al. (2012).

10.4 Cyanobacterial Polysaccharides

Cyanobacteria are a large and diverse group of bacteria that carry out oxygenic photosynthesis. These microorganisms present a considerable morphologic diversity and an unusual capacity for cell differentiation. They have a highly efficient mechanism for adaptation and are able to proliferate in a wide range of habitats (fresh and sea water, soil and harsh environments) (Whitton and Potts, 2000). Morphological groups include unicellular coccoid (spheroidal shape), sometimes attached to a mucilaginous capsule, and multicellular strains, which grow as unbranched or branched filaments. In certain strains, some filament cells can differentiate to carry our specialized functions. In diazotrophic strains, some vegetative cells can differentiate into heterocysts, cells with a thickened cell-wall that contain the nitrogen fixation machinery (Flores and Herrero, 2010). Another type of specialized cells are akinetes, which are larger than vegetative cells with a thicker cell wall, and differentiate from those in response to diverse environmental conditions. Between the cell wall and the mucilaginous layers these cells segregate a new fibrous layer. They have a reduced metabolism and survive harsh life conditions (Kaplan-Levy et al., 2010). Cyanobacteria cytoplasm usually presents corpuscular structures, such as carboxysomes (corpuscles containing ribulose-1,5-biphosphate carboxylase RuBisCO, which carries out CO_2 fixation) and glycogen, cyanophycin, and polyphosphate granules. The molecular photosynthetic machine is located in the thylakoids, which are formed by plasmatic membrane invaginations. The cyanobacterial cover is constituted, as in every Gram-negative bacteria, by a plasmatic and an external membrane, with a peptidoglycan in between.

The polysaccharides produced by cyanobacteria can be divided into three groups: storage, exo-, and cell-envelope polysaccharides. The carbon reserve polysaccharide is glycogen. This homoglucan consists of α-(1,4) linkages with α-(1,6) branches and it bares a close resemblance to green plants amylopectin and animal glycogen. Extraction and determination of glycogen are discussed in Chapter 12, Case Study: Glycogen.

Exopolysaccharide composition and role depend on the cyanobacterial strain and the environmental conditions. Mostly, they are high-molecular-weight heteropolysaccharides that may either be released to the surrounding medium or remain more or less strongly attached to the cell surface (Pereira et al., 2009).

In general terms, regarding the polysaccharides from the cell envelope, cyanobacteria are surrounded by many layers, which play a key role in the interaction between cells and their environment. These layers constitute the cell envelope, which comprises the cell membrane, the cell wall and the external layers. The vegetative-cell wall is mainly constituted by peptidoglycan, proteins and lipopolysaccharides. Its function is to determine and maintain the cell shape and size. The lipopolysaccharide present in cyanobacteria in general is similar to that of bacteria. It is mainly constituted by different sugars. Cyanobacterial cell walls have a structure which bears a close resemblance to that of Gram-negative bacteria; however, the peptiglycan layer found in cyanobacteria is considerably thicker than that of most Gram-negative bacteria (Hoiczyk and Hansel, 2000). Finally, the external layers are made of a lamina, a capsule, and external mucilage. On the other hand, the heterocyst wall is constituted by complex lipids and polysaccharides (Cardemil and Wolk, 1981a,b). The polysaccharides are composed of glucose, mannose, and xylose. To study these polymers, the lipidic layer should be firstly removed by treatment with chloroform/methanol (2/1, v/v) (Cardemil and Wolk, 1976). The resulting aqueous phase contains the polysaccharides whose composition can be analyzed by treatment with different glycosidases (such as glucosidase, mannosidase, and galactosidase).

Experimental Procedures

10.5 Extraction of Mannans

Principle

Mannans are extracted from different plant sources. The biological material is first treated with organic solvents in order to remove lipid compounds, which may interfere in the general procedure. Then, it is extracted with alcohol and mannans precipitate after a copper solution addition. Finally, these polysaccharides are purified by successive precipitations. This procedure may differ according to the plant tissue to be extracted (Matheson, 1990).

Reagents

Biological material: Palm seeds (ivory nuts)	
Acetone/ether	
KOH	7%
Glacial acetic acid	
Ethanol	98%
Fehling's solution	(see below)
HCl	1%

Procedure

Remove the germ and the brownish covering of palm seeds (ivory nuts). Grind endosperms from 100 to 150 g of seeds and extract with acetone/ether in a Soxhlet. The extracted material is sieved and the 13–40 mesh fraction is used for mannan extraction.

Extract the material three times with 7% KOH and leave overnight in a closed vessel with continuous agitation. Neutralize the extract with glacial acetic acid and precipitate mannans with an equal volume of ethanol 98%. Dissolve polysaccharides in 7% KOH. The purification is accomplished after two successive precipitations using Fehling's solution to form copper complexes, followed by the decomposition of these complexes with 1% HCl. The resulting polysaccharides are precipitated with 98% ethanol.

Comments

Fehling's solution is prepared by mixing equal volumes of an aqueous solution of copper(II) sulfate (70 g cupric sulfate pentahydrate per liter of solution) with a clear and colorless solution of aqueous potassium sodium tartrate tetrahydrate (known as Rochelle salt) ($350\ g \cdot L^{-1}$), and a strong alkali (100 g of sodium hydroxide per liter). In this final mixture, aqueous tartrate ions from the dissolved Rochelle salt chelate to Cu^{2+} ions from the dissolved copper(II) sulfate, as bidentate ligands giving a bistartratocuprate(II)4−complex.

The tartarate ions, by complexing copper, prevent the formation of $Cu(OH)_2$ from the reaction of $CuSO_4$, $5H_2O$, and NaOH present in the solution.

10.6 Determination of Mannans

Principle

To carry out studies in land plants, a large number of methods have been developed for the isolation, fractionation, and determination of these polymers (Aspinall, 1982).

Mannans are precipitated and treated with alcohol in order to remove free monosaccharides (such as glucose and fructose). The residue is resuspended in buffer and hydrolyzed with β-(1,4)-mannanase. The released mannose is determined after treatment with hexokinase and mannose-6-phosphate isomerase, followed by glucose-6-phosphate dehydrogenase, to finally measure NADPH appearance by absorbance at 340 nm.

Reagents

Acetate buffer (pH 4.5)	1 M
Tris-HCl buffer (pH 8.0)	1 M
Endo β-(1,4)-mannanase	
Ethanol	80%
Hexokinase	
Mannose-6-phosphate isomerase	
Glucose-6-phosphate dehydrogenase	

Procedure

Resuspend mannan precipitate in 5 mL of 80% (v/v) ethanol. Agitate the tubes in a vortex mixer and add another 5 mL of 80% (v/v) ethanol. Agitate the tubes and centrifuge at 1500 × g for 10 min. Remove carefully the supernatant and leave the material to dry on absorbent paper. Once dry, resuspend the material in 8 mL of acetate buffer (pH 4.5) and agitate the tubes vigorously until the precipitate is completely dispersed. Heat the tubes immediately in a water bath at 100°C for 30 s. Agitate again the tubes in vortex and incubate them in a water bath at 100°C for 4 min in order to make sure that mannans are completely hydrated. Remove the tubes, agitate them in vortex, and put them in a water bath at 40°C. After 5 min, add 20 μL of β-(1,4)-mannanase and vigorously agitate the tubes in vortex for 30 s. Incubate the tubes at 40°C for 60 min agitating them intermittently in vortex (2–3 times). For powder ivory nut samples, a 0.5-mL sample volume is typically enough for the reaction.

Mannose liberated is determined by the hexokinase and mannose-6-phosphate isomerase reaction, followed by glucose-6-phosphate dehydrogenase. The reaction is followed by the formation of NADPH (see Chapter 1: Determination of Carbohydrates Metabolism Molecules).

10.7 Extraction and Determination of Cyanobacterial Insoluble Polysaccharides

Principle

Total insoluble polysaccharides from filamentous nitrogen-fixing cyanobacterial strains are extracted in acid medium from the residue left after glycogen extraction. The released monosaccharides are determined by a colorimetric method (Curatti et al., 2008).

Reagents

Ethanol	80%
HCl	1.8 M
NaOH	4 N

Procedure

Collect cells from 50 mL of a mid-log phase culture of *Anabaena* by centrifugation at 10,000 × g for 10 min. Wash the cell pellet with 80% ethanol, and resuspend in 400 μL of distilled water. Extract glycogen by autoclaving at 120°C for 1 h. Wash the remaining pellet with boiling water, and resuspend it in 500 μL of 1.8 N HCl. Incubate at 100°C for 1 h. Neutralize the reaction with NaOH. Determine the reducing sugar by the Somogyi–Nelson method (see Chapter 1: Determination of Carbohydrates Metabolism Molecules).

10.8 Extraction and Quantification of Cyanobacterial Exopolysaccharides

Principle

The extraction of exopolysaccharides from cyanobacteria (eg, from *Synechocystis* sp. PCC 6803) has been adapted from an isolation method from bacterial exopolysaccharides (Cérantola and Montrozier, 2001; Jittawuttipoka et al., 2013).

Reagents

Ethanol	95%
NaCl	1.5%

Procedure

Harvest the cells by centrifugation at 10,000 × g for 10 min at 4°C. The precipitate obtained is vigorously agitated in vortex for 30 min in a NaCl 1.5% solution and then centrifuged again. Remove the supernatant and heat it at 80°C for 15 min. Centrifuge at 10,000 × g for 10 min to remove any remaining cell. Precipitate exopolysaccharides by adding 4 volumes of ethanol 95%. Let at −80°C for 1 h. Collect exopolysaccharides by centrifugation at 10,000 × g for 15 min, wash with ethanol 95% and let it dry in the air. Resuspend the dry precipitate in sterile water. The total amount of carbohydrates is determined by the anthrone or the phenol-sulfuric acid method (see Chapter 1: Determination of Carbohydrates Metabolism Molecules).

Comments

Alternative exopolysaccharide extraction methods are proposed by Azeredo and Oliveira (1996). These polymers can be precipitated with 1 volume of 4 M NaCl and 2 volumes of 96% isopropanol at 4°C overnight. Polysaccharides are collected by centrifugation at 10,000 × g for 10 min.

10.9 Extraction of Soluble Oligosaccharides from Anabaena and Nostoc Strains

In response to high or fluctuating environmental salinity, cyanobacteria develop different strategies to cope with the rise in the osmotic potential and in cell ion concentration, as a consequence of the loss of water. The two basic strategies are: (1) the enhancement of active ion export systems and (2) the accumulation of organic osmoprotective compounds that are low-molecular-mass molecules with no net charge, known as compatible solutes, as they do not interfere with cellular metabolism (Klähn and Hagemann, 2011). These osmolytes help to reduce the internal osmotic cell potential, maintain membrane integrity, and stabilize proteins. Sucrose accumulation in salt acclimation was reported in many freshwater strains as well as in marine picocyanobacteria. Particularly, filamentous heterocyst-forming strains, belonging to the genera *Anabanea* and *Nostoc*, accumulate sucrose as their main organic osmolyte (Hagemann and Erdmann, 1997) and sucroglucans, compatible solutes recently described (Salerno et al., 2004; Pontis et al., 2007). Sucroglucans are a series of nonreducing oligosaccharides derived from sucrose, where glucose is linked through its hemiacetalic hydroxyl, to the 2 position of the glucose moiety of sucrose. The general structure of these polymers is: (α-D-Glucose-$(1 \rightarrow 2)_n$)-α-D-Glucose-$(1 \rightarrow 2)$-β-D-Fructose. Depending on the strain and cell condition, the continuous series can reach up to 10 hexose residues.

Principle

Sucroglucans are extracted together with other soluble sugars with lightly alkaline water. Thus, the initial steps are identical to those described for sucrose extraction (see Chapter 6: Case Study: Sucrose).

Reagents

Alkaline water (brought to pH ~8 with ammonia solution)	
NaOH	2 M
Acid invertase (from yeast)	
Sodium acetate buffer (pH 4.5)	1 M

Procedure

Grind weighed fresh-cell pellets or freeze-dried material in a mortar under liquid nitrogen. Resuspend the powdered material in alkaline water (brought to pH ~8 with ammonia solution) according to ratio 2 mL water per gram of fresh weight. Heat the suspension at 100°C for 5 min under stirring with a glass rod. Centrifuge at 12,000 × g for 5 min at 4°C. Transfer the supernatant to another tube. Repeat the water extraction procedure twice and pool the supernatants. Freeze-dry the solution containing soluble compounds and resuspend the dry residue in a minimum volume of distilled water.

Incubate in a 50-μL-final volume mixture, an aliquot of the sugar solution, an aliquot of yeast acid invertase, and 5 μL of acetate buffer (pH 4.5) (100 mM final concentration), to hydrolyze sucrose. Destroy monosaccharides by adding alkali up to 0.4 M and heat 10 min in a water bath. Quantify total monosacharides by the Somogyi–Nelson method or fructose content by the Thiobarbituric acid method (see Chapter 6: Determination of Carbohydrates Metabolism Molecules).

Further Reading and References

Allard, B., Tazi, A., 1993. Influence of growth status on composition of extra-cellular polysaccharides from two *Chlamydomonas* species. Phytochemistry. 32, 41–47.

Aspinall, G.O., 1982. Isolation and fractionation of polysaccharides. In: Aspìnall, G.O. (Ed.), The Polysaccharides, vol. 1. Academic Press, New York, pp. 35–131.

Azeredo, J., Oliveira, R., 1996. A new method for precipitating bacterial exopolysaccharides. Biotechnol. Tech. 10, 341–344.

Barclay, W.R., Lewin, R.A., 1985. Microalgal polysaccharide production for the conditioning of agricultural soils. Plant Soil. 88, 159–169.

Brennan, C.S., Blake, D.E., Ellis, P.R., Schofield, J.D., 1996. Effects of guar galactomannan on wheat bread microstructure and on the in vitro and in vivo digestibility of starch in bread. J. Cereal Sci. 24, 151–160.

Cardemil, L., Wolk, C.P., 1976. The polysaccharides from heterocyst and spore envelopes of a blue-green alga. Methylation analysis and structure of the backbones. J. Biol. Chem. 251, 2967–2975.

Cardemil, L., Wolk, C., 1981a. Isolated heterocysts of *Anabaena variabilis* synthesize envelope polysaccharide. Biochim. Biophys. Acta. 674, 265–276.

Cardemil, L., Wolk, C.P., 1981b. Polysaccharides from the envelopes of heterocysts and spores of the blue green algae *Anabaena variabilis* and *Cylindrospermum licheniforme*. J. Phycol. 17, 234–240.

Carlberg, G.E., Percival, E., 1977. The carbohydrates of the green seaweeds *Urospora wormskioldii* and *Codiolum pusillum*. Carbohydrate Res. 57, 223–234.

Cérantola, S., Montrozier, H., 2001. Production in vitro, on different solid culture media, of two distinct exopolysaccharides by a mucoid clinical strain of *Burkholderia cepacia*. FEMS Microbiol. Lett. 202, 129–133.

Craigie, J.S., Leigh, C., 1978. Cans and agars. In: Hellebust, J.A., Craigie, J.S. (Eds.), Handbook of Physiological Methods. Cambridge University Press, Cambridge, pp. 109–131.

Curatti, L., Giarrocco, L.E., Cumino, A.C., Salerno, G.L., 2008. Sucrose synthase is involved in the conversion of sucrose to polysaccharides in filamentous nitrogen-fixing cyanobacteria. Planta. 228, 617–625.

Dey, P.M., 1978. Biochemistry of plant galactomannans. Adv. Carbohydr. Chem. Biochem. 35, 341–376.

Dodgson, K.S., Prince, R.G., 1962. A note on the determination of the ester sulfate content of sulfated polysaccharides. J. Biochem. 84, 106–110.

Edwards, M.E., Dickson, C.A., Chengappa, S., Sidebottom, C., Gidley, M.J., Reid, J.S.G., 1999. Molecular characterization of a membrane-bound galactosyltransferase of plant cell wall matrix polysaccharide biosynthesis. Plant J. 19, 691–697.

Flores, E., Herrero, A., 2010. Compartmentalized function through cell differentiation in filamentous cyanobacteria. Nature Rev. Microbiol. 8, 39–50.

Frei, E., Preston, R.D., 1968. Non-cellulosic structural polysaccharides in algal cell walls. III. Mannan in siphoneous green algae. Proc. R Soc. Lond. B Biol. Sci. 169, 127–145.

Hagemann, M., Erdmann, N., 1997. Environmental stresses. In: Rai, A.K. (Ed.), Cyanobacterial Nitrogen Metabolism and Environmental Biotechnology. Springer-Verlag, Narosa Publishing House, New Delhi, pp. 155–221.

Hoiczyk, E., Hansel, A., 2000. Cyanobacterial cell walls: news from an unusual prokaryotic envelope. J. Bacteriol. 18, 1191–1199.

Jabbar-Mian, A.J., Percival, E., 1973. Carbohydrates of the brown seaweeds *Himanthalia lorea*, *Bifurcaria bifurcata* and *Padina pavonia*: part I. Extraction and fractionation. Carbohydr. Res. 26, 133–146.

Jittawuttipoka, T., Planchon, M., Spalla, O., Benzerara, K., Guyot, F., Cassier-Chauvat, C., et al., 2013. Multidisciplinary evidences that Synechocystis PCC 6803 exopolysaccharides operate in cell sedimentation and protection against salt and metal stresses. PLoS One. 8 (2), e55564.

Jones, A.S., Letham, D.S., 1954. A submicro method for the estimation of sulphur. Chem. Ind. 662–663, London.

Kaplan-Levy, R.N., Hadas, O., Summers, M.L., Rücker, J., Sukenik, A., 2010. Akinetes: dormant cells of cyanobacteria, in dormancy and resistance in harsh environments. In: Lubzens, E., Cerda, J., Clark, M. (Eds.), Topics in Current Genetics, vol. 21. Springer-Verlag, Berlin, Heidelberg.

Klähn, S., Hagemann, M., 2011. Compatible solute biosynthesis in cyanobacteria. Environ. Microbiol. 13, 551–562.

Liepman, A.H., Nairn, C.J., Willats, W.G.T., Sørensen, I., Roberts, A.W., Keegstra, K., 2007. Functional genomic analysis supports conservation of function among cellulose synthase-like a gene family members and suggests diverse roles of mannans in plants. Plant Phys. 143, 1881–1893.

Mackie, W., Preston, R.D., 1968. The occurrence of mannan microfibrils in the green algae *Codium fragile* and *Acetabularia crenulata*. Planta. 79, 249–253.

Maeda, Y., Awano, T., Takabe, K., Fujita, M., 2000. Immunolocalization of glucomannans in the cell wall of differentiating tracheids in *Chamaecyparis obtusa*. Protoplasma. 213, 148–156.

Matheson, N.K., 1990. Mannose-based polysaccharides. In: Dey, P.M., Harborne, J.B. (Eds.), Methods in Plant Biochemistry. Academic Press, New York, pp. 371–413.

Meier, H., Reid, J.S.G., 1982. Reserve polysaccharides other than starch in higher plants. In: Loewua, F.A., Tanner, W. (Eds.), Encyclopedia of Plant Physiology, New Series, vol. 13A. Springer, New York, pp. 418–471.

Mills, A.R., Timell, T.E., 1963. Constitution of three hemicelluloses from the wood of engelmann spruce (*Picea engelmanni*). Can. J. Chem. 41, 1389–1395.

Painter, T.J., 1983. Algal polysaccharydes. In: Aspinall, G.O. (Ed.), The Polysaccharides. Academic Press, New York, pp. 195–285.

Percival, E., 1978. Sulfated polysaccharides metabolized by the marine chlorophyceae. ACS Symp. Ser. 77, 203–212.

Percival, E., McDowell, R.H., 1981. Algal walls composition and biosynthesis. In: Loewua, F.A., Tanner, W. (Eds.), Encyclopedia the Plant Physiology, New Series, Vol. 13B. Springer, New York, pp. 277–316.

Percival, E.E., Venegas, J., Weigel, H., 1983. Carbohydrates of the brown seaweed *Lessonia nigrescens*. Phytochemistry. 22, 1429–1432.

Pereira, S., Zille, A., Micheletti, E., Moradas-Ferreira, P., De Philippis, R., Tamagnini, P., 2009. Complexity of cyanobacterial exopolysaccharides: composition, structures, inducing factors and putative genes involved in their biosynthesis and assembly. FEMS Microbiol. Rev. 33, 917–941.

Pontis, H.G., Vargas, W.A., Salerno, G.L., 2007. Structural characterization of the members of a polymer series, compatible solutes in *Anabaena* cells exposed to salt stress. Plant Sci. 172, 29–35.

Salerno, G.L., Porchia, A.C., Vargas, W.A., Abdian, P.L., 2004. Fructose-containing oligosaccharides: novel compatible solutes in *Anabaena* cells exposed to salt stress. Plant Sci. 167, 1003–1008.

Scott, J.E., 1960. Aliphatic ammonium salts in the assay of acidic polysaccharides from tissues. In: Glick, D. (Ed.), Methods in Biochemical Analysis, 8. Interscience, New York, p. 163.

Stephen, A.M., 1983. Other plant polysaccharides. In: Aspinal, G.O. (Ed.), The Polysaccharides, vol. 2. Academic Press, New York, pp. 97–193.

Sui, Z., Gizaw, Y., BeMiller, J.N., 2012. Extraction of polysaccharides from a species of *Chlorella*. Carbohydr. Polym. 90, 1–7.

Whitton, B.A., Potts, M., 2000. Introduction to the cyanobacteria. In: Whitton, B.A., Potts, M. (Eds.), The Ecology of Cyanobacteria. Kluwer Academic Publishers, Dordrecht, pp. 1–11.

CHAPTER 11

Case Study: Starch

Chapter Outline

11.1 Introduction

Starch is the main form of carbon storage in plants and green algae. It is widely distributed among plant species and it is present in most tissues. It consists of glucose polymers arranged into a semicrystalline structure (starch granule). It is an important end product of the photosynthetic process synthesized in the leaves or in other green tissues during the day and mobilized at night. But, it is also produced in amyloplasts of storage tissues such as seeds, tubers, roots, fruits, and pollen grains. Starch present in plant chloroplasts is found in a transient form, while that accumulated in storage-organ amyloplasts is in a more stable form (Whistler and Daniel, 1984; Preiss, 1988). In algae, while starch is found in the cytosol of all glaucophytes (single-cell freshwater algae containing a plastid called cyanelle) and red algae (Rhodophyceae), it is found in the plastids of all green (Chloroplastida) algae (Ball et al., 2011).

Methods for Analysis of Carbohydrate Metabolism in Photosynthetic Organisms.
DOI: http://dx.doi.org/10.1016/B978-0-12-803396-8.00011-9

Properties and structure

Starch granules are composed of two major α-glucans: amylose (essentially a linear polyglucan) and amylopectin (a branched polyglucan) and consist of granular rings of alternating semicrystalline and amorphous regions in layers. Amylose is a product of the condensation of D-glucopyranoses by α-(1→4) glucosydic bonds, which sets long glucose linear chains of variable degrees of polymerization and molecular weights up to 10^6 daltons. Thus, amylose is an α-D-(1→4)-glucan whose repetitive unit is α-maltose. It easily acquires a helical three-dimensional conformation, in which every helical turn consists of six glucose molecules. On the inside of the helix there are only hydrogen atoms, while the hydroxyl groups are located at the outside of the helix. Most amylose molecules contain a small number of α-(1→6) branches (approximately one branch per 1000 residues) and makes up about 25–30% of starch. This proportion may vary considerably with the plant species (Martin and Smith, 1995). High-amylose corn starches, which are commercially available, have approximately between 50% and 75% of amylose. Amylopectin, the major component of starch (c.70%), is a branched molecule which may contain up to 10^6 glucosyl residues, with an average chain length of 15–25 α-D-(1→4)-linked glucosyl units that are linked by α-D-(1→6) branch linkages. Even though the exact molecular-weight value of the large macromolecule that the chains form is difficult to achieve, it was calculated as in the order of 10^7–10^8 (Aberle et al., 1994; Buléon et al., 1998). Waxy starches are exclusively constituted by amylopectin.

The starch granules are structurally complex and their properties predominantly depend on the amylase:amylopectin ratio, the length and degree of branching of the amylose and amylopectin chains, and the presence of minor components (eg, phosporylated residues, mineral ions, lipids, and some proteins) (Avigad and Dey, 1997). Starch granules can be easily recognized after tissue staining with iodine or by electron microscopy.

11.2 Starch Biosynthesis and Degradation

The enzymatic reactions leading to the production of starch are similar in both the chloroplasts and the amyloplasts, and involve the enzymes: ADP-glucose pyrophosphorylase (AGPase, EC 2.7.7.27), starch synthase (EC 2.4.1.21), and starch branching enzyme (SBE, EC 2.4.1.28), which catalyze the following reactions:

$$\text{Glucose-1-phosphate} + \text{ATP} \xleftrightarrow{\text{AGPase}} \text{ADP-glucose} + \text{Pyrophosphate (PPi)}$$

$$\text{ADP-glucose} + (1\rightarrow4)\text{-}(\alpha\text{-D-glucosyl})_n \xleftrightarrow{\text{Starch synthase}} (1\rightarrow4)\text{-}(\alpha\text{-D-glucosyl})_{n+1} + \text{ADP}$$

$$\text{Linear-}(1\rightarrow4)\ \alpha\text{-glucan} \xleftrightarrow{\text{SBE}} \text{Branched } (1\rightarrow6)\ (1\rightarrow4)\ \alpha\text{-glucan}$$

The first step in starch biosynthesis is ADP-glucose synthesis from glucose-1-phosphate and ATP catalyzed by the specific pyrophosphorylase (AGPase). This reaction is the major rate-controlling step in the starch biosynthesis pathway, and, consequently, it is subjected to several regulatory mechanisms (such as allosteric regulation by metabolites, transcriptional regulation, and redox regulation by thioredoxin) (Ballicora et al., 2004). In the second step, ADP-glucose is the glucose donor for α-glucan elongation in the reaction catalyzed by the enzyme starch synthase. The glucose transfer from ADP-glucose to the nonreducing end of the α-glucan is responsible for the elongation of both amylose and amylopectin chains. Different starch synthase isoforms have been found in almost all plant tissues and in green algae. They can be grouped into: (1) granule-bound starch synthases (GBSS) that are involved in amylose synthesis and are located within the granule matrix and (2) starch synthases that can be found either in the granules or in the stromal fractions. The distribution of these starch synthases vary between species, tissues, and the developmental stage (Ball and Morell, 2003). Additionally, the starch-branching enzyme, which acts as a complex together with the soluble starch synthase, catalyzes the hydrolysis of an α-(1→4)-linkage and the subsequent transfer of an α-(1→4)-glucan to form an α-(1→6)-branching point generating branches in chains with α-(1→6) linkages and producing amylopectin nonreducing ends. Therefore, linear and branched glucans are synthesized by the sequential actions of starch synthase and starch-branching enzyme activities.

Starch degradation in both, photosynthetic and nonphotosynthetic tissues has different mechanisms and dynamics that still remain unclear. In general, it occurs in three phases: reduction of the granules to soluble maltodextrins, debranching, and degradation of larger maltodextrins to glucose and glucose-1-phosphate. Different enzymes participate in starch degradation in either phosphorolytic or hydrolytic cleavage reactions (Zeeman et al., 2010). Phosphorolytic degradation involves at least three enzymes: starch phosphorylases (EC 2.4.1.1), starch-debranching enzymes (EC 3.2.1.68, isoamylase-type, and EC 3.2.1.41, pullulanase-type), and glucosyltransferases:

$$(1\rightarrow4)\text{-}(\alpha\text{-D-glucosyl})_n + \text{Orthophosphate (Pi)} \xleftrightarrow{\text{Starch phosphorylase}} (1\rightarrow4)\text{-}(\alpha\text{-D-glucosyl})_{n-1} + \alpha\text{-D-glucose-1-phosphate}$$

$$\text{Branched } (1\rightarrow6)\ (1\rightarrow4)\ \alpha\text{-glucan} \xleftrightarrow{\text{Starch-debranching enzyme}} \text{Linear-}(1\rightarrow4)\ \alpha\text{-glucan}$$

$$(1\rightarrow4)\text{-}(\alpha\text{-D-glucosyl})_m + (1\rightarrow4)\text{-}(\alpha\text{-D-glucosyl})_n \xleftrightarrow{\text{Glucosyltransferase } (D\text{-enzyme})} (1\rightarrow4)\text{-}(\alpha\text{-D-glucosyl})_{m+n-1} + \text{Glucose}$$

Table 11.1: Starch hydrolyzing enzymes

Enzyme	Action	Bond Specificity	Final Product
Phosphorylase	Exo	α-(1,4)-Glucosyl	Glucose-1-phosphate
α-Amylase (i)	Endo	α-(1,4)-Glucosyl	Linear and branched dextrins
β-Amylase	Exo	α-(1,4)-Glucosyl	Maltose y β-dextrins
Amyloglucosidase	Exo	α-(1,4)-Glucosyl and α-(1,6)-glucosyl	Glucose
Pullulanase	Endo	α-(1,6)-Glucosyl with interval > G_3	Linear α-(1,4)-glucan chains
Isoamylase	Endo	α-(1,6)-Glucosyl	Linear α-(1,4)-glucan chains
α-Glucosidase	Exo	α-(1,2), α-(1,3) and α-(1,4)-Glucosyl	Glucose

Source: Republished with permission from Morrison, W.R., Karkalas, J., 1990. Starch. In: Dey, P.M. (Ed.), Carbohydrates, Methods in Plant Biochemistry, vol. 2, pp. 323–352. (Morrison and Karkalas, 1990)

Starch can also be degraded by α-amylase (EC 3.2.1.1) that catalyzes the internal cleavage of glucosyl bonds, producing shorter glucans (dextrines) and some glucose and maltose, and by β-amylase (EC 3.2.1.2), which yields maltose and short-chain glucans. Finally, maltose, dextrins, and short-chain glucans are degraded by α-glucosidase (EC 3.2.1.20).

$$(1\rightarrow 4)\text{-}(\alpha\text{-D-glucosyl})_n \xrightarrow{\alpha\text{-Amylase}} (1\rightarrow 4)\text{-}(\alpha\text{-D-glucosyl})_x(1\rightarrow 4)\text{-}(\alpha\text{-D-glucosyl})_y \quad (n \geq 3;\ x+y=n)$$

$$(\alpha\text{-D-glucosyl})_n + H_2O \xrightarrow{\beta\text{-Amylase}} \text{Maltose} + (\alpha\text{-D-glucosyl})_{n-2}$$

$$(1\rightarrow 4)\text{-}(\alpha\text{-D-glucosyl})_n + H_2O \xrightarrow{\alpha\text{-Glucosidase}} (1\rightarrow 4)\text{-}(\alpha\text{-D-glucosyl})_{n-1} + \text{Glucose}$$

The end products of the enzymes that hydrolyze starch polysaccharides are shown in Table 11.1.

Experimental Procedures

11.3 Starch Extraction

11.3.1 Extraction from Leaves

Principle

After the extraction of soluble sugars, starch is solubilized from the insoluble precipitate. The method is adapted from procedures described for extracting starch from *Arabidopsis* and wheat leaves (Strand et al., 1999; Trevanion, 2000).

Reagents

Ethanol	80%
Hepes-NaOH buffer (pH 7.5)	4 mM
Alternatively, alkaline water (approximately pH 8.0)	

Procedure

Weigh the harvested leaves, and immediately freeze them in liquid nitrogen. Grind the leaves to fine powder in a mortar and pestle under liquid nitrogen. Extract soluble sugars with 2 mL of water per gram of fresh weight (see Chapter 6: Case Study: Sucrose, Section 6.3) or alternatively, with 80% ethanol containing 4 mM Hepes-KOH (pH 7.5) at 80°C for 30 min. Centrifuge for 15 min at 11,000 × g. Discard the supernatant. Resuspend the pellet in 80% ethanol-Hepes, pH 7.5. Heat at 80°C again for 30 min. Repeat the extraction twice: once with 50% ethanol-Hepes-KOH (pH 7.5), and once with only 4 mM Hepes-KOH (pH 7.5). Combine supernatants for soluble sugar assays (Stitt et al., 1989). Resuspend the pellet in 0.5 mL of distilled water and autoclave the sample at 121°C for 3 h. Starch determination is achieved in an aliquot of the autoclaved suspension.

Comments

A similar procedure can be used for starch extraction from root seedlings. Soluble metabolites can be also extracted with 50 mM KOH at 70–80°C for 20 min. Insoluble material is removed by centrifugation (13,000 × g for 15 min at 4°C) and starch extraction is achieved by autoclaving the precipitated residue at 121°C.

11.3.2 Extraction from Potato Tubers

Principle

This is a mild method that allows extracting starch together with active granule-bound starch synthase (Edwards et al., 1995). Potato tubers from actively growing plants are the biological material source.

Reagents

Tris-HCl (pH 7.5)	0.5 M
Neutralized EDTA	0.2 M
Dithiotreitol (DTT)	0.1 M
Sodium metabisulphite	

Extraction medium: Prepare 1 L of extraction medium by mixing 100 mL of 0.5 M Tris-HCl (pH 7.5), 5 mL of EDTA, 10 mL of 0.1 M DTT, and 10 mg of sodium metabisulphite.

Procedure

Cut into small pieces freshly harvested potato tubers (c.50–100 g). Homogenize immediately in a blender with 100 mL of ice-cold 50 mM Tris-HCl buffer (pH 7.5) containing 1 mM EDTA and 1 mM DTT and 10 $mg \cdot L^{-1}$ of sodium metabisulphite. Filtrate the homogenate through two layers of muslin. Allow to settle. Discard the supernatant and resuspend the pellet in 200 mL of extraction medium. Centrifuge at $1000 \times g$ for 10 min at 4°C. Repeat the resuspension and centrifugation four times. Resuspend the pellet in 50 mL of cold acetone (at −20°C). Allow to settle for 2 min. Discard the supernatant and wash the remaining pellet twice with acetone. Dry the final pellet, which is immediately used for assaying granule-bound starch synthase activity. Store the remainder at −20°C.

Comments

Alternatively, tubers can be frozen at once in liquid nitrogen and stored at −80°C prior to use. Samples are extracted three times in 10-mL lots of 80% ethanol at 80°C for 10 min to remove soluble saccharides. The extracted material is homogenized and autoclaved for 2 h. Starch content of the resulting solution is determined by incubation with hydrolytic enzymes (see Section 11.4).

11.3.3 Extraction of Intact Starch Granules

Principle

Three different procedures to extract intact starch granules are described. The first method is applied to extract starches from tubers to be used in physicochemical studies (aspects of granules and gelatinization characteristics). The starch quality allows comparing granule properties (Yusuph et al., 2003).

The second method described is generally used to extract intact granules from plant leaves and from some unicellular green algae (Zeeman et al., 1998). Finally, the third extraction procedure is employed to extract pure native starch from *Ostreococcus tauri* (a marine picophytoeukaryote species that synthesizes a unique starch granule at the center of its single chloroplast). Starch yields through this purification procedure are greater than 80% (Ral et al., 2004).

(A) Extraction from potato tubers

Reagents

Sodium thiosulphate
Sodium chloride
Prepare a solution 1% (w/v) sodium chloride and 1% (w/v) sodium thiosulphate
CsCl 80%

Procedure

Wash the potato tubers in cold water. Homogenize the tubers in a mini blender in cold sodium thiosulphate:sodium chloride solution. Filter the material through a muslin cloth. Concentrate the starch by centrifugation at 1500 × g for 5 min. Then centrifuge the concentrate through 80% (w/v) CsCl at 30,000 × g for 20 min at 15°C (Tester and Morrison, 1990). Wash extensively the starch obtained with cold water. After each wash, concentrate the starch by centrifugation at 1500 × g for 5 min. Finally, rinse twice with acetone and allow to air dry to obtain a dry powder (Yusuph et al., 2003).

(B) Extraction from *Arabidopsis thaliana* leaves

Reagents

3-(N-morpholino)propanesulfonic acid (MOPS) buffer (pH 7.2)	0.5 M
EDTA	0.5 M
Ethanediol	
Sodium dodecyl sulfate (SDS)	

Homogenization buffer (100 mM MOPS buffer, 5 mM EDTA, and 10% (v/v) ethanediol): To prepare 100 mL of homogenization buffer, mix 20 mL of MOPS buffer (pH 7.2) 0.5 M, 1 mL neutralized Na-EDTA, and 10 mL of ethanediol. The SDS-homogenization buffer contains 5% (w/v) of SDS.

Procedure

Weigh leaves harvested at the end of the photoperiod (c.15–20 g of fresh weight). Wash the leaves and homogenize them using a Polytron blender adding 5 volumes of the homogeneization buffer. Filter the homogenate through two layers of Miracloth and a 20 μm nylon mesh. Centrifuge at 3000 × g for 10 min at 4°C. Resuspend the pellet in 30 mL of the homogenization buffer containing 0.5% (w/v) of SDS. Wash the precipitated starch twice more with SDS-homogenization buffer, and then six times with 30 mL of deionized water. Confirm intactness of the granules by scanning electron microscopy (Zeeman et al., 1998).

(C) Extraction from *Ostreococcus tauri* cells

Reagents

Pluronic	0.2%
Tris-acetate buffer (pH 7.5)	10 mM containing 1 mM EDTA
Percoll	90%

Procedure

Harvest cells cultured in a nitrogen-limited media for 4 days and under continuous illumination. Centrifuge cells at 10,000 × g for 20 min with 0.2% Pluronic (a nonionic detergent). Discard the supernatant and resuspend the pellet in 300 μL of 10 mM Tris-acetate buffer (pH 7.5) containing 1 mM EDTA. Disrupt the cells (at a density of 10^8 cells · mL^{-1}) by sonication keeping the sample in an ice-water bath (10 cycles at 40 W, with 20-s pauses between cycles). Centrifuge the lysate at 10,000 × g for 15 min. Resuspend the pellet in 90% Percoll (adding 1 mL of Percoll per liter of culture). Centrifuge at 10,000 × g for 30 min. Collect the starch pellet from the formed gradient and resuspend it in 1 mL of 90% Percoll. Centrifuge again at 10,000 × g for 30 min. Rinse the starch pellet in sterile distilled water. Recover the purified starch by centrifugation at 10,000 × g for 30 min, and keep dry at 4°C for immediate use or freeze-dry for subsequent analysis.

Comments

An alternative method can be used to prepare native starch from unicellular algae (such as *Chlamydomonas reinhardtii*) (Delrue et al., 1992). Cells are ruptured by sonication and the lysate is centrifuged at 2000 × g for 20 min. The pellet is washed in 10 mM Tris-ClH buffer (pH 8.0) containing 1 mM EDTA. The resuspended pellet (in 1 mL of the same buffer per 10^8 cells) is passed twice through a Percoll gradient (9 mL of Percoll per mL of crude starch pellet). The purified starch pellet is washed and centrifuged at 2000 × g in distilled water for subsequent immediate analysis or freeze-dried.

11.4 Starch Determination

11.4.1 Determination After Enzymatic Hydrolysis

Principle

Enzymatic methods are very specific, sensitive, and reproducible. α-Amyloglucosidase (also known as glucoamylase, EC 3.2.1.3) catalyzes the sequential hydrolysis of α-(1→4)- and α-(1→6)-D-glucosidic bonds in polysaccharides (such as maltosides (maltose, maltotriose, maltotetraose, maltopentaose, etc.), dextrins, starch, glycogen), releasing β-D-glucose molecules from the nonreducing terminal end. Also the enzyme α-amylase (EC 3.2.1.1) cleaves internal α-(1→4) glycosidic linkages in starch to produce glucose, maltose, or dextrins. Due to the high specificity of these two enzymes, amyloglucosidase and α-amylase are important tools to be used to quantitatively determine starch (or glycogen). Some protocols use only amyloglucosidase and other protocols use both enzymes.

$$\text{Starch} + (n-1)\,H_2O \xrightarrow{\alpha\text{-Amyloglucosidase and }\alpha\text{-amylase}} n\ D\text{-glucose}$$

The released glucose molecules can be quantified by either of the methods described in Chapter 1, Determination of Carbohydrates Metabolism Molecules.

Reagents

Sodium acetate buffer (pH 4.8)	50 mM
α-Amyloglucosidase (from *Aspergillus niger*)	
α-Amylase (from *Bacillus licheniformis*)	
Soluble starch (from potato)	1%

Enzyme solution: Dissolve 5 mg of protein in 1 mL of 10 mM sodium acetate buffer (pH 4.8). If the dissolution is incomplete, centrifuge and use the supernatant.

Standard curve: To dissolve the starch in distilled water for the standard curve, heat between 60°C and 80°C for approximately 15 min.

Procedure

For starch cleavage, add a 50-μL aliquot of the autoclaved suspension (soluble starch sample) to 450 μL of 50 mM sodium-acetate incubation buffer (pH 4.8), containing 28 units of amyloglucosidase and 36 units of α-amylase. Incubate the sample at 37°C for 16 h. Centrifuge at 11,000 × g for 15 min and assay the supernatant for glucose (see Chapter 1: Determination of Carbohydrates Metabolism Molecules).

Comments

Starch cleavage can be carried out by incubating with only α-amyloglucosidase. An alternative protocol is described as follows. Glucose or maltodextrins present in the sample can be removed with ethanol. Add enough ethanol up to 40% final concentration. Agitate and then centrifuge at 10,000 × g for 15 min at 4°C. To extract starch, remove the supernatant and keep the precipitate. Suspend the precipitate in 500 μL of acetate buffer (pH 4.8). Heat for 120 min at 130°C. Centrifuge the sample at 12,000 × g for 15 min at 4°C. Remove the supernatant and incubate with α-amyloglucosidase. For each 100 μL of reaction mixture (containing 1% (w/v) of soluble starch), add 10 μL of acetate buffer and 10 μL of enzyme solution (α-amyloglucosidase). Incubate the reaction mixture at 55°C for 1 h. Stop the reaction by heating at 100°C for 2 min. Centrifuge at 10,000 × g for 15 min at 4°C. Glucose released is measured as described in Chapter 1, Determination of Carbohydrates Metabolism Molecules.

11.4.2 Determination by a Colorimetric Method

The total amount of starch can be determined without polysaccharide hydrolysis by the method developed by Krisman (1962) to estimate glycogen. The method, based on the colorimetric determination of the complex formed between glycogen and iodine in presence of calcium chloride as a color stabilizer, is described in Chapter 12, Case Study: Glycogen, Section 12.4.2.

11.5 ADP-Glucose Pyrophosphorylase Activity Assays

ADP-glucose pyrophosphorylase (AGPase) catalyzes the reversible synthesis of ADP-glucose and pyrophosphate (PPi) from ATP and glucose-1-phosphate. To assay AGPase activity, methods with different sensitivity, experimental difficulty, and cost have been developed. Their characteristics, and pros and cons have been analyzed and discussed by Fusari et al. (2006).

AGPase activity can be determined in the pyrophosphorolysis of ADP-glucose direction determining glucose-1-phosphate in a NAD-linked phosphoglucomutase/glucose-6-phosphate dehydrogenase coupled system (Plaxton and Preiss, 1987; Ball et al., 1991) or by the formation of [^{32}P]ATP from ^{32}PPi (Ghosh and Preiss, 1966; Fu et al., 1998; Ballicora et al., 2000), as described in Chapter 15, Case Study: Nucleotide Sugars. Also, AGPase can be assayed in the ADP-glucose synthesis direction using [^{14}C]-glucose-1-phosphate and ATP as substrates (Plaxton and Preiss, 1987). Fusari et al. (2006) developed a relatively simple, highly sensitive, accurate, and reliable colorimetric method to assay AGPase in both of the reaction directions by quantifying inorganic orthophosphate released from the specific hydrolysis of the enzyme activity products.

In Chapter 15, Case Study: Nucleotide Sugars, the most frequent methods used to assay plant AGPases are described.

11.6 Starch Synthase Activity Assays

11.6.1 Granule-Bound Starch Synthase Activity Assays

Assay method A

Principle

This procedure permits a separation of the products of the granule-bound starch synthase reaction from the enzyme by simple centrifugation. Starch synthase activity is determined by the amount of ADP formed in the reaction (when the incubation is carried out with unlabeled ADP-glucose) or by measuring the incorporation of [^{14}C]-glucose into the starch granules (Cardini and Frydman, 1966).

Reagents

Glycine buffer (pH 8.6)	100 mM
EDTA	25 mM
ADP-[^{14}C]-glucose (specific activity 7.4 GBq · mol^{-1})	20 mM
or Unlabeled ADP-glucose	20 mM
Ethanol	50% (v/v)
Methanol	75% (v/v)

Procedure

In a 20-μL total volume, mix 4 μL of glycine buffer, 10 μL of ADP-glucose (ADP-[^{14}C]-glucose (approximately 50,000 cpm)), and 4 mg of enzyme preparation. Incubate at 37°C for 30–90 min, depending on the starch synthase source. Incubation can also be carried out using unlabeled ADP-glucose.

After incubation, granule-bound starch synthase activity is determined by quantifying the radioactivity incorporated (1) or by measuring the amount of ADP formed (2).

(1) Determination of label incorporation into the starch granules

Stop the reaction by adding 500 μL of 50% ethanol. Centrifuge the suspension at 10,000 × g for 10 min at 4°C. Wash the pellet four times with 500 μL of 50% ethanol. Suspend the final pellet in 400–500 μL of distilled water. Disperse the starch by heating at 100°C for 10 min. Count the radioactivity incorporated in a scintillation counter after the addition of liquid scintillation solution.

(2) Determination of ADP formation

Quantify the amount of ADP formed by the pyruvate kinase procedure (see Chapter 1: Determination of Carbohydrates Metabolism Molecules, Section 1.6.1). Add the pyruvate kinase and the phosphoenol pyruvate without inactivating the enzyme. Note that if the reaction mixture is heated, a starch paste is formed and the further enzymatic reaction is difficult to proceed.

Assay method B

Principle

Starch synthase activity is measured from pure native starch granules purified following the Percoll method (Delrue et al., 1992) (see above, Section 11.3.3(C)), using labeled ADP-glucose (Tenorio et al., 2003).

Reagents

Hepes-NaOH buffer (pH 8.5)	50 mM
Methanol-KCl solution	75% (v/v) methanol, 1% (w/v) KCl
Methanol/KCl solution	70% (v/v) methanol, 1% (w/v) KCl

Stock solutions to prepare the resuspension buffer

Tricine buffer (pH 8.5)	1 M
Potassium acetate	0.5 M
Dithiotreitol (DTT)	100 mM
EDTA	100 mM
ADP-glucose	100 mM
ADP-[^{14}C]-glucose	

Procedure

Wash the extracted starch granules three times in 50 mM Hepes-NaOH buffer (pH 7.5). Resuspend the granules in 200 μL of the resuspension buffer (100 mM Tricine buffer (pH 8.5) containing 25 mM potassium acetate, 10 mM DTT, 5 mM EDTA, and 10 mM ADP-[^{14}C]-glucose (7.4 GBq · mol^{-1})). Separate 100 μL and heat at 100°C for 3 min (time = 0 of the assay). Incubate the remaining 100 μL for 1 h at 30°C. Stop the reaction by heating at 100°C for 3 min. Wash the starch granules three times with a solution of 70% methanol, 1% KCl (Denyer et al., 1995). Count the radioactivity incorporated into the starch granules in a scintillation counter.

11.6.2 Soluble Starch Synthase Activity Assay

Soluble starch synthase activity is determined as described by Ghosh and Preiss (1965) with modifications (Jenner et al., 1994; Cao et al., 1999; Szydlowski et al., 2009).

Reagents

Tricine buffer (pH 8.0)	1 M
Potassium acetate	0.5 M
EDTA	100 mM
Sodium citrate	1 M
Bovine serum albumin (BSA)	10 mg · mL^{-1}
Dithiotreitol (DTT)	100 mM
Potato amylopectin	20 mg · mL^{-1}
ADP-[^{14}C]glucose (3.7 GBq.mol–1)	20 mM
Methanol	75%
Methanol-KCl solution	75% (v/v) methanol containing 1% (w/v) KCl

Procedure

In a total volume of 100 μL, mix 10 μL of Tricine buffer (pH 8.0), 5 μL of potassium acetate, 5 μL of EDTA, 5 μL of DTT, 50 μL of sodium citrate, 5 μL of BSA, 5 μL of amylopectin, and 10 μL of enzyme. Initiate the reaction by the addition of 5 μL of ADP-[^{14}C]-glucose (3.7 GBq · mol^{-1}). Incubate at 30°C for 30 min. Stop the reaction by boiling at 100°C for 2 min. Add 1.5 mL of 75% methanol containing 1% KCl. Centrifuge at 10,000 × g 10 min at 4°C, Rinse the pellet twice with 1 mL of 75% methanol/1% KCl solution. Dry at room temperature for 30 min. Resuspend the pellet in 300 mL of distilled water. Count in a scintillation counter after the addition of counting solution (Delvallé et al., 2005).

Comments

From the whole oligosaccharides series tested in Preiss's work, only those derived from the maltodextrins series functioned as primers or acceptors. An enzymatic activity unit is defined as the amount of enzyme that transforms 1 nmol of ADP-[^{14}C]-glucose per min in the material soluble in methanol under the fixed experimental condition.

An alternative protocol to the methanol precipitation (Jenner et al., 1994) is described as follows. Prepare small columns filling 1-mL pipettes with a suspension of Dowex 1-X8 anion exchange resin 200–400 mesh chloride form (25 g of resin suspended in 100 mL of water). Centrifuge at 50 × g for 2 min. After terminating the enzyme incubation by heating at 100°C, transfer the reaction mixture to the previously prepared Dowex 1-X8 column supported over a scintillation vial. Centrifuge at 50 × g for 2 min. Rinse the reaction tube twice with 50 μL of distilled water, applying the washing to the column each time. Centrifuge after each addition. Determine the radioactivity in the column eluate by liquid scintillation spectrometry.

11.7 Starch Branching Enzyme Activity Assay

Principle

The assay is based on the stimulation of α-glucan formation from glucose-1-phosphate catalyzed by rabbit muscle phosphorylase (Boyer and Preiss, 1978) modified to be nonradioactive by Fisher et al. (1996). In the absence of the SBE, long amylose chains are formed by de novo synthesis at a very slow rate, since the concentration of end groups with which phosphorylase reacts remains very low. In the presence of branching enzyme, the number of end groups increases and the phosphorylase reaction proceeds at a faster rate. Thus, the amount of orthophosphate (Pi) released in this system is an indirect measure of SBE activity.

Reagents

Hepes-NaOH (pH 7.0)	1 M
AMP	10 mM
Glucose-1-phosphate	500 mM
Crystalline rabbit muscule phosphorylase a	

Procedure

In a 50-μL total volume, mix 5 μL of buffer Hepes-NaOH (pH 7.0), 5 μL of 10 mM AMP, 10 μg of crystalline rabbit-muscle phosphorylase a, an aliquot of enzyme preparation, and 5 μL of glucose-1-phosphate (initiate the reaction by addition of glucose-1-phosphate). Incubate at 30°C for 60–90 min. Stop the reaction by heating at 100°C for 1 min. Released Pi is determined using a colorimetric method (such as the Fiske–Subbarow procedure described in Chapter 1: Determination of Carbohydrates Metabolism Molecules). SBE activity is expressed in terms of the micromoles of Pi formed (and, therefore, of glucosyl residues polymerized) per min per mg protein.

11.8 Phosphorylase Activity Assay

The activity of starch phosphorylase is currently measured either by determining Pi formation (with a colorimetric method) or by the incorporation of [^{14}C]-glucose from [^{14}C]-glucose-1-phosphate.

Principle

Phosphorylase activity can be assayed by colorimetric or spectrophotometrically methods, determining the polysaccharide degradation or formation. The colorimetric method is based on the estimation of Pi, which is released from glucose-1-phosphate after the addition of a glucose residue to the polysaccharide present in the reaction mixture. On the other hand, the spectrophotometric method is based on the estimation of glucose-1-phsophate released from the polysaccharide cleavage in the presence of inorganic phosphate and determining the increase in optical density at 340 nm due to NADPH formation by coupling phosphoglucomutase and glucose-6-phosphate-dehydrogenase (see Chapter 1: Determination of Carbohydrates Metabolism Molecules).

Reagents

Glucose-1-phosphate pH 6.3	0.1 M
Glucose-1,6-diphosphate (pH 6.3)	0.1 mM
Amylopectin	2.5%
Citrate buffer (pH 6.3)	0.5 M
Trichloroacetic acid	5%
Sodium acetate	0.2 M

Procedure

In 0.1-mL total volume, mix 30 μL of 2.5% amylopectin, 20 μL of 0.5 M citrate buffer (pH 6.3), 10 μL of glucose-1-phosphate, 10 μL of glucose-1,6-diphosphate, and an aliquot of enzyme preparation, completing the volume with water. Initiate the reaction with the addition of the enzyme to the mixture pre-incubated at 30°C for 5 min. Incubate the complete mixture for 5 min at 30°C. Stop the reaction by adding 50 μL of trichloroacetic acid 5%. Add 0.1 mL of 0.2 M sodium acetate to adjust the pH over 4 to avoid glucose-1-phosphate hydrolysis. The amount of Pi formed in the enzymatic reaction can be determined by the Fiske–Subbarow method (see Chapter 1: Determination of Carbohydrates Metabolism Molecules).

11.9 Amylase Activity Assay

Principle

α-Amylase, the main enzyme involved in starch breakdown during germination, hydrolyzes α-(1→4)bonds in amylose and amylopectin, releasing fragments that can be further broken down by β-amylase and α-glucosidase. The specific α-amylase activity measurement in plant crude extracts is difficult because the presence of β-amylases may interfere. The starch-azure assay seems to be specific for α-amylase and fairly satisfactory for most tissues. Insoluble starch azure (obtained from potato starch derivatized with Remazol Brilliant Blue (RBB)) is a chromogenic substrate specific for α-amylase that allows the estimation of α-amylase activity as a function of the color intensity produced by the release of soluble fragments of starch linked to the RBB dye (Doehlert and Duke, 1983; Hirasawa, 1989; Sanwo and DeMason, 1992).

Reagents

Starch azure solution:	2% (w/v) starch azure suspended in 0.10 M citrate buffer (pH 5.5) and 3 mM $CaCl_2$
Trichloracetic acid (TCA)	20% (w/v)

Note: To prepare the starch azure solution, heat the suspension slowly to boiling while stirring continuously. Stir the azure suspension continuously until use. The solution can be stored at 4°C, and reheated to boiling under stirring before being reused.

Procedure

Mix 4 mL of starch azure suspension and 1 mL of enzyme preparation. Incubate at 30°C for 20 min. Mix thoroughly and transfer 1-mL aliquot to a tube containing 1 mL of 20%. Centrifuge at about 3000 × g for 20 min to precipitate the unreacted starch azure. Determine the absorbance of the supernatant at 595 nm. Prepare blanks by taking 1 mL aliquot from the reaction mixture at time zero of incubation. Absorbance values are converted to μmol of RBB released with the use of a standard curve.

Comments

To specifically determine α-amylase activity in the presence β-amylases, selective inactivation of β-amylases by heating at 70°C for 20 min, $HgCl_2$ treatment, and the use of starch azure (α-amylase specific substrate) have been proposed (Doehlert and Duke, 1983).

On the other hand, in leaf starch degradation, while α-amylases were shown not to play a crucial role, β-amylases are essential to release maltose from the nonreducing ends of α-(1→4)-glucan chains with a degree of polymerization of 4 or greater (Weise et al., 2005; Sparla et al., 2006). Thus, amylase activity may be ascribed to the action of β-amylase.

Further Reading and References

Aberle, T., Burchard, W., Vorwerg, W., Radosta, S., 1994. Conformational contributions of amylose and amylopectin to the structural properties of starches from various sources. Starch. 46, 329–335.

Avigad, G., Dey, P.M., 1997. Carbohydrate metabolism: storage carbohydrates. In: Dey, P.M., Harborne, J.B. (Eds.), Plant Biochemistry. Academic Press, London, pp. 143–204.

Ball, S.G., Morell, M.K., 2003. From bacterial glycogen to starch: understanding the biogenesis of the plant starch granule. Annu. Rev. Plant Biol. 54, 207–233.

Ball, S., Marianne, T., Dirick, L., Fresnoy, M., Delrue, B., Decq, A.A., 1991. *Chlamydomonas reinhardtii* low-starch mutant is defective for 3-phosphoglycerate activation and orthophosphate inhibition of ADP-glucose pyrophosphorylase. Planta. 185, 17–26.

Ball, S.G., Colleoni, C., Cenci, U., Raj, J.N., Tirtiaux, C., 2011. The evolution of glycogen and starch metabolism in eukaryotes gives molecular clues to understand the establishment of plastid endosymbiosis. J. Exp. Bot. 62, 1775–1801.

Ballicora, M.A., Frueauf, J.B., Fu, Y., Schürmann, P., Preiss, J., 2000. Activation of the potato tuber ADP-glucose pyrophosphorylase by thioredoxin. J. Biol. Chem. 275, 1315–1320.

Ballicora, M.A., Iglesias, A.A., Preiss, J., 2004. ADP-glucose pyrophosphorylase: a regulatory enzyme for plant starch synthesis. Photosynth. Res. 79, 1–24.

Boyer, C.D., Preiss, J., 1978. Multiple forms of starch branching enzyme of maize: evidence for independent genetic control. Biochem. Biophys. Res. Commun. 80, 169–175.

Buléon, A., Colonna, P., Planchot, V., Ball, S., 1998. Starch granules: structure and biosynthesis. Int. J. Biol. Macromol. 23, 85–112.

Cao, H., Imparl-Radosevich, J., Guan, H., Keeling, P.L., James, M.G., Myers, A.M., 1999. Identification of the soluble starch synthase activities of maize endosperm. Plant Physiol. 120, 205–215.

Cardini, C.E., Frydman, R.B., 1966. ADP-glucose: α-1, 4-glucan glucosyltransferases (starch synthetases and related enzymes) from plants. In: Neufeld, E.F., Ginsburg, V. (Eds.), Complex Carbohydrates. Methods in Enzymology, vol. VIII. Academic Press, New York and London, pp. 387–394.

Delvallé, D., Dumez, S., Wattebled, F., Roldán, I., Planchot, V., Berbezy, P., et al., 2005. Soluble starch synthase I: a major determinant for the synthesis of amylopectin in *Arabidopsis thaliana* leaves. Plant J. 43, 398–412.

Denyer, K., Hylton, C.M., Jenner, C.E., Smith, A.M., 1995. Identification of multiple isoforms of soluble and granule-bound starch synthase in developing wheat endosperm. Planta. 196, 256–265.

Delrue, B., Fontaine, T., Routier, F., Decq, A., Wieruszeski, J.M., Van Den Koornhuyse, N., et al., 1992. Waxy *Chlamydomonas reinhardtii*: monocellular algal mutants defective in amylose biosynthesis and granule-bound starch synthase activity accumulate a structurally modified amylopectin. J. Bacteriol. 174, 3612–3620.

Doehlert, D.C., Duke, S.H., 1983. Specific determination of α-amylase activity in crude plant extracts containing β-amylase. Plant Physiol. 71, 229–234.

Edwards, A., Marshall, J., Sidebottom, C., Visser, R.G., Smith, A.M., Martin, C., 1995. Biochemical and molecular characterization of a novel starch synthase from potato tubers. Plant J. 8, 283–294.

Fisher, D.K., Gao, M., Kim, K.N., Boyer, C.D., Guiltinan, M.J., 1996. Allelic analysis of the maize amylose-extender locus suggests that independent genes encode starch-branching enzymes IIa and IIb. Plant Physiol. 110, 611–619.

Fu, Y., Ballicora, M.A., Leykam, J.F., Preiss, J., 1998. Mechanism of reductive activation of potato tuber ADP-glucose pyrophosphorylase. J. Biol. Chem. 273, 25045–25052.

Fusari, C., Demonte, A.M., Figueroa, C.M., Aleanzi, M., Iglesias, A.A., 2006. A colorimetric method for the assay of ADP-glucose pyrophosphorylase. Anal Biochem. 352, 145–147.

Ghosh, H.P., Preiss, J., 1965. Biosynthesis of starch in spinach chloroplasts. Biochemistry. 4, 1354–1361.

Ghosh, H.P., Preiss, J., 1966. Adenosine diphosphate glucose pyrophosphorylase. A regulatory enzyme in the biosynthesis of starch in spinach leaf chloroplasts. J. Biol. Chem. 241, 4491–4504.

Hirasawa, E., 1989. Auxins induce α-amylase activity in pea cotyledons. Plant Physiol. 91, 484–486.

Jenner, C.F., Denyer, K., Hawker, J.S., 1994. Caution on the use of the generally accepted methanol precipitation technique for the assay of soluble starch synthase in crude extracts of plant tissues. Aust. J. Plant Physiol. 21, 17–22.

Krisman, C.R., 1962. A method for the colorimetric estimation of glycogen with iodine. Anal. Biochem. 4, 17–23.

Martin, C., Smith, A.M., 1995. Starch biosynthesis. Plant Cell. 7, 971–985.

Morrison, W.R., Karkalas, J., 1990. Starch. In: Dey, P.M. (Ed.), Carbohydrates, Methods in Plant Biochemistry, vol. 2. Academic Press, London, pp. 323–352.

Plaxton, W.C., Preiss, J., 1987. Purification and properties of nonproteolytic degraded ADPglucose pyrophosphorylase from maize endosperm. Plant Physiol. 83, 105–112.

Preiss, J., 1988. Biosynthesis of starch and its regulation. In: Preiss, J. (Ed.), The Biochemistry of Plants: Carbohydrates, structure and function, vol. 14. Academic Press, New York, pp. 181–254.

Ral, J.P., Derelle, E., Ferraz, C., Wattebled, F., Farinas, B., Corellou, F., et al., 2004. Starch division and partitioning. A mechanism for granule propagation and maintenance in the picophytoplanktonic green alga *Ostreococcus tauri*. Plant Physiol. 136, 3333–3340.

Sanwo, M.M., DeMason, D.A., 1992. Characteristics of α-amylase during germination of two high-sugar sweet corn cultivars of *Zea mays* L. Plant Physiol. 99, 1184–1192.

Sparla, F., Costa, A., Lo Schiavo, F., Pupillo, P., Trost, P., 2006. Redox regulation of a novel plastid-targeted ß-amylase of *Arabidopsis*. Plant Physiol. 141, 840–850.

Stitt, M., Lilley, R.M., Gerhardt, R., Heldt, H.W., 1989. Metabolite levels in specific cells and subcellular compartments of plant leaves. In: Fleischer, S., Fleischer, B. (Eds.), Methods in Enzymology: Biomembranes, vol. 174. Academic Press, Amsterdam, pp. 518–552.

Strand, A., Hurry, V., Henkes, S., Huner, N., Gustafsson, P., Gardeström, P., et al., 1999. Acclimation of *Arabidopsis* leaves developing at low temperatures. Increasing cytoplasmic volume accompanies increased activities of enzymes in the Calvin cycle and in the sucrose-biosynthesis pathway. Plant Physiol. 119, 1387–1398.

Szydlowski, N., Ragel, P., Raynaud, S., Lucas, M.M., Roldán, I., Montero, M., et al., 2009. Starch granule initiation in *Arabidopsis* requires the presence of either class IV or class III starch synthases. Plant Cell. 21, 2443–2457.

Tenorio, G., Orea, A., Romero, J.M., Mérida, A., 2003. Oscillation of mRNA level and activity of granule-bound starch synthase I in *Arabidopsis* leaves during the day/night cycle. Plant Mol. Biol. 51, 949–958.

Tester, R.F., Morrison, W.R., 1990. Swelling and gelatinization of cereal starches. I. Effects of amylopectin, amylose and lipids. Cereal Chem. 67, 551–557.

Trevanion, S.J., 2000. Photosynthetic carbohydrate metabolism in wheat (*Triticum aestivum* L.) leaves: optimization of methods for determination of fructose 2,6-bisphosphate. J. Exp. Bot. 51, 1037–1045.

Weise, S.E., Kim, K.S., Stewart, R.P., Sharkey, T.D., 2005. β-Maltose is the metabolically active anomer of maltose during transitory starch degradation. Plant Physiol. 137, 756–761.

Whistler, R.L., Daniel, J.L., 1984. Molecular structure of starch. In: Whistler, R.L., BeMiller, J.N., Paschal, E.F. (Eds.), Starch: Chemistry and Technology, second ed. Academic Press, Orlando, pp. 153–182.

Yusuph, M., Tester, R.F., Ansell, R., Snape, C.E., 2003. Composition and properties of starches extracted from tubers of different potato varieties grown under the same environmental conditions. Food Chem. 82, 283–289.

Zeeman, S., Northrop, F., Smith, A.M., Rees, T.A., 1998. A starch-accumulating mutant of *Arabidopsis thaliana* deficient in a chloroplastic starch hydrolyzing enzyme. Plant J. 15, 357–365.

Zeeman, S.C., Kossmann, J., Smith, A.M., 2010. Starch: its metabolism, evolution, and biotechnological modification in plants. Ann. Rev. Plant Biol. 61, 209–234.

CHAPTER 12

Case Study: Glycogen

Chapter Outline

12.1 Introduction

Glycogen is the most widespread form of carbon storage in nature: it is found in Archaea, Bacteria, and Eukaryotes (Ball et al., 2011). Similarly to other bacteria, glycogen is the major storage polysaccharide in cyanobacteria. These microorganisms are among the most diverse groups of prokaryotes capable of performing oxygen-evolving photosynthesis like plants do. Cyanobacterial strains exhibit a wide range of morphologies (from unicellular to various multicellular organizations) and a remarkable capacity to colonize the most different habitats (marine, freshwater, or terrestrial ecosystems) and to adapt to environmental changes (Whitton, 1992).

In cyanobacteria, glycogen biosynthesis is closely connected with the photosynthetic process. The carbon assimilation reactions produce glucose-1-phosphate, the precursor of glycogen, which is stored in granules. The polysaccharide accumulates in the light period and provides carbon and respiratory substrates in the darkness, acting as a dynamic reserve. As well as in heterotophic bacteria, in cyanobacteria many environmental factors, such as pH, low and high temperatures, and salinity, have been proved to regulate glycogen accumulation (Preiss, 1984; Wilson et al., 2010). Glycogen is the main source of carbon

Methods for Analysis of Carbohydrate Metabolism in Photosynthetic Organisms.
DOI: http://dx.doi.org/10.1016/B978-0-12-803396-8.00012-0

and energy readily available to cope with adverse conditions (Smith, 1982; Page-Sharp et al., 1998; Suzuki et al., 2010; Zilliges, 2014). For example, the ability of some strains to acclimate to salinity changes involves the synthesis of compatible solutes (such as glucosylglycerol and sucrose) which is related to glycogen accumulation (Hagemann, 2011; Kolman et al., 2015). However, light and macronutrient (such as nitrogen) are the main environmental factors that control glycogen levels (Yoo et al., 2007; Gründel et al., 2012). It is also important to highlight the crucial role of glycogen in diazotrophic strains to achieve nitrogen fixation and its connection with sucrose metabolism (Ernst and Boger, 1985; Cumino et al., 2007; Curatti et al., 2008; Vargas et al., 2011).

Even though most cyanobacteria accumulate glycogen, some strains have distinct α-polyglucans, named as semi-amylopectin. In terms of chain length distribution, molecular size, and length of the most abundant α-1,4-chain, semi-amylopectin is an intermediate polyglucan between rice amylopectin and typical cyanobacterial glycogen. It was also reported that there is a strain that lacks amylose-type components in its α-polyglucans (Nakamura et al., 2005; Suzuki et al., 2013).

A general overview of glycogen determination as well as general methods to assay the main enzymes involved in its metabolism are described below.

Properties and structure

Cyanobacterial glycogen, as well as that present in animals, insects, and eubacteria, is a water-soluble polymer of α-(1→4)- and α-(1→6)-linked glucose residues. Its structure may be similar to starch amylopectin, although the latter is much less branched than glycogen. The α-(1→6) branches account for 7–10% of the linkages and are evenly distributed within the glycogen particle. The characteristic branching pattern consists of each glucan chain (except the outer unbranched chains) supporting two branches, which permits a spherical growth of the particle generating tiers. Consequently, there is an increase in the density of chains in each tier leading to a progressively more crowded structure towards the periphery. In vivo, glycogen particles are present in the form of granules and a single granule consists of 12 tiers (42 nm maximal diameter), composed of 55,000 glucose residues (Ball et al., 2011). On the other hand, branching also allows an abundance of nonreducing glucose residues (the union sites of glycogen phosphorylase and glycogen synthase enzymes), which facilitate both polymer synthesis and degradation. The particles are completely hydrosoluble which implies that soluble glycogen degrading enzymes have a ready access to the polyglucan leading to a rapid mobilization.

12.2 Synthesis and Degradation of Glycogen

Similarly to starch biosynthesis (see Chapter 11: Case Study: Starch), the metabolic pathway conducting the glycogen synthesis involves three enzymatic steps, which are likely to be conserved among cyanobacteria, from analysis of annotated genome sequences

(Beck et al., 2012; Zilliges, 2014). The first step is the production of the glucose donor (ADP-glucose) by the enzyme ADP-glucose pyrophosphorylase (AGPase, EC 2.7.7.27), according to the following reaction:

$$\text{Glucose-1-phosphate} + \text{ATP} \xleftrightarrow{\text{AGPase}} \text{ADP-glucose} + \text{Pyrophosphate (PPi)}$$

In bacteria, ADP-glucose is the activated-glucose form used in glycogen synthesis, while in all Eukaryotes (except those belonging to Archaeplastida), glycogen synthesis is carried out from UDP-glucose (Ball et al., 2011).

In the second step, a glucosyl group is transferred from the nucleotide sugar to the nonreducing end of a pre-existent linear α-(1→4)-glucan through an elongation reaction catalyzed by glycogen synthase (GSase, EC 2.4.1.21):

$$\text{ADP-glucose} + (1\rightarrow4)\text{-}(\alpha\text{-D-glucosyl})_n \xleftrightarrow{\text{GSase}} (1\rightarrow4)\text{-}(\alpha\text{-D-glucosyl})_{n+1} + \text{ADP}$$

To reach the final particle structure, the branching enzyme(s) (GBE, EC 2.4.1.18), and the debranching enzyme(s) (DBE, EC 3.2.1) catalyze the addition and removal of α-(1→6)-linked ramifications, respectively (Yoo et al., 2002; Suzuki et al., 2007).

The glycogen biosynthesis route (similarly to the plant starch pathway) is regulated at the level of ADP-glucose synthesis, with AGPase activity being highly modulated (Iglesias et al., 1991; Gómez Casati et al., 1999; Díaz-Troya et al., 2014).

Glycogen degradation in cyanobacteria is not as well understood as is its synthesis (Zilliges, 2014) and different enzymes are required. It is initiated by glycogen phosphorylase (GlgP, EC 2.4.1.1), an enzyme that cleaves α-(1→4) glycosylic linkages into glucose-1-phosphate from the available nonreducing ends in the presence of orthophosphate. Phosphorylase stops four glucose residues away from an α-(1→6) branch.

$$(1\rightarrow4)\text{-}(\alpha\text{-D-glucosyl})_n + \text{Orthophosphate (Pi)} \xleftrightarrow{\text{GlgP}} (1\rightarrow4)\text{-}(\alpha\text{-D-glucosyl})_{n-1} + \alpha\text{-D-glucose-1-phosphate}$$

In a second step, the resulting limit dextrins that contain mostly external chains of 4 glucose residues, are readily degraded by the debranching enzyme (EC 3.2.1.68), producing maltotetraose (Ball and Morell, 2003), according to the following general reaction:

$$\text{[External four-glucose chain]-glycogen} \xrightarrow{\text{Debranching enzyme}} \text{Glycogen} + \text{Maltotetraose}$$

The glycogen product of this reaction has a longer outer chain and becomes a substrate for the phosphorylase. Maltotetraose is further degraded by α-(1→4)-glucanotransferase (EC 2.4.1.25) and phosphorylases.

Experimental Procedures

12.3 Glycogen Extraction

Principle

Glycogen can be extracted from cyanobacterial cells harvested by centrifugation using different procedures: (A) extraction with hot potassium hydroxide after removing soluble sugars from the cell precipitate (modified from Hassid and Abraham, 1957); (B) a glass-bead extraction method (Yoo et al., 2007); and (C) extraction under pressure (autoclaving) at high temperature (Curatti et al., 2008).

Method A

Reagents

KOH	33% (w/v)
Ethanol	96% (v/v)
Alkaline water (brought to pH ~8.0 with ammonia solution)	

Procedure

Add alkaline water to precipitated-weighed cells (2 mL of water per gram of fresh weight). Transfer the suspension to a corex tube and heat at 100°C for 5 min, under continuous stirring with a glass rod. Cool the tube and centrifuge for 5 min at 10,000 × g at 4°C. Remove the supernatant (containing soluble sugars). Use the precipitate for glycogen extraction. Wash the precipitate with alkaline water three times, heating at 100°C for 5 min and centrifuging each time. Dissolve the precipitate in 33% KOH (add 900 μL of KOH to 100 mg of fresh weight). Incubate at 100°C for 2 h and centrifuge at 2000 × g for 5 min at 4°C. Remove the supernatant and add ice-cold ethanol up to a final concentration of ~75% (v/v). Incubate on ice for at least 2 h. Centrifuge at 10,000 × g for 10 min at 4°C. Wash the glycogen pellet twice with ethanol (70–96%). Dry the pellet in a speed-vac. Dissolve the precipitate in 500 μL of water (or 50–100 mM sodium acetate buffer pH 4.5–5.8) per 100 mg of fresh weight. Heat the solution at 100°C for 1 min in order to help glycogen dissolution. Store at −20°C.

Method B

Reagents

100–150 μm glass beads	
Ethanol	96% (v/v)

Procedure

Disperse the cell pellet in 1.5 mL of water. Add an equal volume of glass beads with that of the cells, and break the cells with a mini-bead beater. Centrifuge at 10,000 × g for 15 min. Add 5 volumes of ethanol to precipitate glycogen from the supernatant. Centrifuge at 8000 × g for 20 min to recover the glycogen in the precipitate. Air-dry the precipitate. Dissolve in water or buffer as was indicated in Method A. Store at −20°C.

Method C

Reagents

Ethanol 80% (v/v)

Procedure

Wash the cell precipitate with 2 volumes of 80% ethanol with that of cells. Centrifuge at 8000 × g for 20 min. Resuspend the precipitate in distilled water (400 μL of water per 100 mg fresh weight). Autoclave at 120°C for 1 h. Centrifuge at 8000 × g for 20 min to recover the glycogen in the precipitate. Dissolve in water or buffer as was indicated in Method A. Store at −20°C.

Comments

The g-force value and time used to collect the cells depend on the cyanobacterial strain. Centrifugation at 6000 × g for 15 min is used for unicellular strains. For different filamentous strains, higher and different g-forces and time of centrifugation are usually used.

12.4 Glycogen Determination

12.4.1 Determination After Enzymatic Hydrolysis

Glycogen is enzymatically hydrolyzed to glucose by treatment with α-amyloglucosidase (eg, adding 2 mg · mL^{-1} amyloglucosidase from *Aspergillus niger*) at pH 4.5–4.8, at 60°C for 2–4 h, and the released monosaccharide is further quantified (see Chapter 1: Determination of Carbohydrates Metabolism Molecules). A similar procedure is described for starch determination in Chapter 11, Case Study: Starch.

12.4.2 Determination by a Colorimetric Method

Principle

Analytical methods based on the color that glycogen gives with iodine are not employed because they lack sensitivity and are temperature-affected. Krisman (1962) developed an accurate procedure for the colorimetric determination of the iodine-glycogen complex in the presence of a salt that makes it more sensible and stable. An adaptation of that method is described below.

Reagents

Iodine/iodide solution:	0.26 g of I_2 and 2.6 g of KI are dissolved in 10 mL of water
$CaCl_2$ solution:	Saturated solution (at room temperature)
Iodine reagent preparation:	Mix 130 mL of $CaCl_2$ solution with 500 μL of I_2/KI solution This solution is stable for a week in a dark bottle at 4°C

Procedure

Add 2.6 mL of the iodine reagent to 0.4 mL of glycogen sample. Mix the solution. In the presence of glycogen, a yellowish brown color is obtained. Measure the absorbance at 460 nm. Absorbance is proportional to glycogen concentration (up to 1 $mg \cdot mL^{-1}$).

12.5 ADP-Glucose Pyrophosphorylase Activity Assays

Different methods to assay AGPase activity in both directions (ADP-glucose synthesis or pyrophosphorolysis) are described in Chapter 15, Case Study: Nucleotide Sugars and commented on in Chapter 11, Case Study: Starch.

12.6 Glycogen Synthase Activity Assay

Principle

Glycogen synthase activity is determined measuring either ADP produced in the reaction or by quantifying the incorporation of $[^{14}C]$-glucose to the glycogen molecule.

$$\text{ADP-glucose} + (1 \rightarrow 4)\text{-}(\alpha\text{-D-glucosyl})_n \xleftrightarrow{\text{GSase}} (1 \rightarrow 4)\text{-}(\alpha\text{-D-glucosyl})_{n+1} + \text{ADP}$$

12.6.1 Determination of ADP

Reagents

Glycil-glycine buffer (pH 8.5)	0.1 M containing 0.025 M EDTA
ADP-glucose	25 mM
Glycogen solution	2.5%

Note: Total glycogen dissolution takes a few minutes. Shake the solution before using.

Procedure

In a 75-μL total volume, mix 5 μL of 1 M glycil-glycine buffer (pH 8.5) containing 0.025 M of EDTA, 25 μL of 2.5% glycogen, 10 μL of ADP-glucose, and an aliquot of the enzyme preparation. Incubate at 30°C for 30 min. Stop the reaction by heating at 100°C for 2 min. Determine the amount of ADP formed in the reaction mixture by the measurement of pyruvate using pyruvate kinase (see Chapter 1: Determination of Carbohydrates Metabolism Molecules, Section 1.16).

12.6.2 Determination of Radioactivity Incorporated to Glycogen

Principle

Glycogen synthase activity assay is based on determining the incorporation of labeled glucose residues into α-glucan molecules used as primers (Fox et al., 1976).

Reagents

Bicine buffer (pH 8.5)	1 M
Potassium acetate	1 M
Magnesium acetate	5 mM
ADP-[^{14}C]-glucose (500 cpm · nmol^{-1})	
Glutathione reduced form (GSH)	100 mM
Bovine serum albumin (BSA)	
Methanol-KCl solution	75% methanol containing 1% KCl
Bacterial glycogen	

Procedure

In a 100-μL total volume, mix 140 nmol of ADP-[^{14}C]-glucose (500 cpm.nmol^{-1}), 5 μL 1 M bicine buffer (pH 8.5), 5 μL of potassium acetate, 10 μL of GSH, 5 μL of magnesium acetate, 50 μg of BSA, 0.3 mg of glycogen, and an aliquot of the enzyme preparation. Incubate at 30°C for 15 min. Stop the reaction by heating at 100°C for 1 min. To a 50 μL-aliquot add 0.1 mL of glycogen solution (10 mg · mL^{-1}) and 2 mL of 75% methanol containing 1% KCl. Cool in an ice-water bath for 5 min. Centrifuge the glycogen precipitated for 5 min at 2000 × g. Redissolve the precipitate in 0.2 mL of water and reprecipitate with 75% methanol-KCl solution and centrifuge. Wash the precipitate once with 2 mL of methanol-KCl solution, and centrifuge. Dissolve in 1.0 mL of water. Add scintillation liquid and count in a scintillation spectrometer (Preiss and Greenberg, 1965).

Comments

In this procedure, it is feasible to use glycogen of animal origin (rabbit liver glycogen) for the determination of cyanobacterial enzyme activity.

12.7 Glycogen Phosphorylase Activity Assay

Principle

Glycogen phosphorylase activity is assayed in the glycogenolysis direction following the method of Andersen and Westergaard (2002) with modifications (Fu and Xu, 2006). The procedure is based on determining glucose-1-phosphate formed by coupling phosphoglucomutase and glucose-6-phosphate dehydrogenase, and measuring NADPH formation at 340 nm.

$$\text{Glucose-1-phosphate} \xleftrightarrow{\text{Phosphoglucomutase}} \text{Glucose-6-phosphate}$$

$$\text{Glucose-6-phosphate} + \text{NADP}^+ \xrightarrow{\text{Glucose-6-phosphate dehydrogenase}} \text{6-Phosphogluconate} + 2\ \text{NADPH}$$

Reagents

Phosphate buffer (pH 6.8)	180 mM KH_2PO_4 and 270 mM Na_2HPO_4
$MgCl_2$	150 mM
EDTA	10 mM
$NADP^+$	34 mM
Glucose-1,6-diphosphate	0.4 mM
Glucose-6-phosphate dehydrogenase (from yeast)	
Ghosphoglucomutase (from yeast)	
glycogen	$100\ mg \cdot mL^{-1}$

Procedure

Prepare the reaction mixture in a 1-mL cuvette with 1-cm light path. In a 1-mL total volume, mix 100 μL of phosphate buffer (pH 6.8), 100 μL of 150 mM $MgCl_2$, 10 μL of 10 mM EDTA, an aliquot of the enzyme extract, 10 μL of 34 mM $NADP^+$, 10 μL of 0.4 mM glucose-1,6-diphosphate, 6 units of glucose-6-phosphate dehydrogenase, and 0.8 units of phosphoglucomutase. Place the cuvette in a suitable thermostatted spectrophotometer (at 25°C). Start the reaction with the addition of 20 μL of $100\ mg \cdot mL^{-1}$ glycogen. Run a control without glycogen. Register the increase in absorbance at 340 nm every minute until constant is observed. Calculate the μmoles of glucose-1-phosphate in the mixture, dividing the increase in absorbance at 340 nm by 6.22.

References

Andersen, B., Westergaard, N., 2002. The effect of glucose on the potency of two distinct glycogen phosphorylase inhibitors. Biochem. J. 367, 443–450.

Ball, S.G., Morell, M.K., 2003. From bacterial glycogen to starch: understanding the biogenesis of the plant starch granule. Annu. Rev. Plant Biol. 54, 207–233.

Ball, S.G., Colleoni, C., Cenci, U., Raj, J.N., Tirtiaux, C., 2011. The evolution of glycogen and starch metabolism in eukaryotes gives molecular clues to understand the establishment of plastid endosymbiosis. J. Exp. Bot. 62, 1775–1801.

Beck, C., Knoop, H., Axmann, I.M., Steuer, R., 2012. The diversity of cyanobacterial metabolism: genome analysis of multiple phototrophic microorganisms. BMC Genomics. 13, 56.

Cumino, A.C., Marcozzi, C., Barreiro, R., Salerno, G.L., 2007. Carbon cycling in *Anabaena* sp. PCC 7120. Sucrose synthesis in the heterocysts and possible role in nitrogen fixation. Plant Physiol. 143, 1385–1397.

Curatti, L., Giarrocco, L.E., Cumino, A.C., Salerno, G.L., 2008. Sucrose synthase is involved in the conversion of sucrose to polysaccharides in filamentous nitrogen-fixing cyanobacteria. Planta. 228, 617–625.

Díaz-Troya, S., López-Maury, L., Sánchez-Riego, A.M., Roldán, M., Florencio, F.J., 2014. Redox regulation of glycogen biosynthesis in the cyanobacterium *Synechocystis* sp. PCC 6803: analysis of the AGP and glycogen synthases. Molec. Plant. 7, 87–100.

Ernst, A., Boger, P., 1985. Glycogen accumulation and the induction of nitrogenase activity in the heterocyst-forming cyanobacterium *Anabaena variabilis*. J. Gen. Microbiol. 131, 3147–3153.

Fox, J., Kawaguchi, K., Greenberg, E., Preiss, J., 1976. Biosynthesis of bacterial glycogen. Purification and properties of the *Escherichia coli* B ADPglucose: 1, 4-α-D-glucan 4-α-glucosyltransferase. Biochemistry. 15, 849–857.

Fu, J.F., Xu, X., 2006. The functional divergence of two *glgP* homologues in *Synechocystis* sp. PCC 6803. FEMS Microbiol. Lett. 260, 201–209.

Gründel, M., Scheunemann, R., Lockau, W., Zilliges, Y., 2012. Impaired glycogen synthesis causes metabolic overflow reactions and affects stress responses in the cyanobacterium *Synechocystis* sp. PCC 6803. Microbiology. 158, 3032–3043.

Gómez Casati, D.F., Aon, M.A., Iglesias, A.A., 1999. Ultrasensitive glycogen synthesis in cyanobacteria. FEBS Lett. 446, 117–121.

Hagemann, M., 2011. Molecular biology of cyanobacterial salt acclimation. FEMS Microbiol. Rev. 2011 (35), 87–123.

Hassid, W.Z., Abraham, S., 1957. Determination of glycogen by modified Pflüger method. In: Colowick, S.P., Kaplan, N.O. (Eds.), Methods in Enzymology, vol. III. Academic Press, Inc., New York, p. 34.

Iglesias, A.A., Kakefuda, G., Preiss, J., 1991. Regulatory and structural properties of the cyanobacterial ADP-glucose pyrophosphorylase. Plant Physiol. 97, 1187–1195.

Kolman, M.A., Nishi, C.N., Perez-Cenci, M., Salerno, G.L., 2015. Sucrose in cyanobacteria: from a salt-response molecule to play a key role in nitrogen fixation. Life. 5, 102–126. Available from: http://dx.doi.org/10.3390/life5010102.

Krisman, C.R., 1962. A method for the colorimetric estimation of glycogen with iodine. Anal. Biochem. 4, 17–23.

Nakamura, Y., Takahashi, J., Sakurai, A., Inaba, Y., Suzuki, E., Nihei, S., et al., 2005. Some cyanobacteria synthesize semi-amylopectin type α-polyglucans instead of glycogen. Plant Cell Physiol. 46, 539–545.

Page-Sharp, M., Behm, C.A., Smith, G.C., 1998. Cyanophycin and glycogen synthesis in a cyanobacterial *Scytonema* species in response to salt stess. FEMS Microbiol. Lett. 160, 11–15.

Preiss, J., 1984. Bacterial glycogen synthesis and its regulation. Annu. Rev. Microbiol. 38, 419–458.

Preiss, J., Greenberg, E., 1965. Biosynthesis of bacterial glycogen. III. The adenosine diphosphate-glucose: α-4-glucosyl transferase of *Escherichia coli* B. Biochemistry. 4, 2328–2334.

Smith, A.M., 1982. Modes of cyanobacterial carbon metabolism. In: Carr, N.G., Whitton, D.A. (Eds.), The Biology of Cyanobacteria. Blackwell Scientific Publications, UK, pp. 47–85. Chapter 3.

Suzuki, E., Ohkawa, H., Moriya, K., Matsubara, T., Nagaike, Y., Iwasaki, I., et al., 2010. Carbohydrate metabolism in mutants of the cyanobacterium *Synechococcus elongatus* PCC 7942 defective in glycogen synthesis. Appl. Environ. Microbiol. 76, 3153–3159.

Suzuki, E., Onoda, M., Colleoni, C., Ball, S., Fujita, N., Nakamura, Y., 2013. Physicochemical variation of cyanobacterial starch, the insoluble α-Glucans in cyanobacteria. Plant Cell Physiol. 54, 465–473.

Suzuki, E., Umeda, K., Nihei, S., Moriya, K., Ohkawa, H., Fujiwara, S., et al., 2007. Role of the GlgX protein in glycogen metabolism of the cyanobacterium, *Synechococcus elongatus* PCC 7942. Biochim. Biophys. Acta. 1770, 763–773.

Vargas, W.A., Nishi, C.N., Giarrocco, L.E., Salerno, G.L., 2011. Differential roles of alkaline/neutral invertases in *Nostoc* sp. PCC 7120: Inv-B isoform is essential for diazotrophic growth. Planta. 233, 153–162.

Whitton, B.A., 1992. Diversity, ecology and taxonomy of the cyanobacteria. In: Mann, N.H., Carr, N.G. (Eds.), Photosynthetic Prokaryotes. Plenum Press, New York, pp. 1–51.

Wilson, W.A., Roach, P.J., Montero, M., Baroja-Fernández, E., Muñóz, F.J., Eydallin, G., et al., 2010. Regulation of glycogen metabolism in yeast and bacteria. FEMS Microbiol. Rev. 34, 952–985.

Yoo, S.H., Spalding, M.H., Jane, J.-l, 2002. Characterization of cyanobacterial glycogen isolated from the wild type and from a mutant lacking of branching enzyme. Carbohyd. Res. 337, 2195–2203.

Yoo, S.H., Kappel, C., Spalding, M., Jane, J.L., 2007. Effects of growth condition on the structure of glycogen produced in cyanobacterium *Synechocystis* sp. PCC6803. Int. J. Biol. Macromolec. 40, 498–504.

Zilliges, Y., 2014. Glycogen, a dynamic cellular sink and reservoir for carbon. In: Flores, E., Herrero, A. (Eds.), The Cell Biology of Cyanobacteria. Caister Academic Press, Norfolk, UK, pp. 189–210. Chapter 8.

CHAPTER 13

Case Study: Cellulose

Chapter Outline

13.1 Introduction

Cellulose is the most abundant biopolymer on earth. It is synthesized by green plants, most algae, many eubacteria (including Gram-positive anaerobic bacteria, purple bacteria, such as species belonging to the genera *Acetobacter*, *Alcaligenes*, *Agrobacterium*, and *Rhizobium*), and cyanobacteria (Brown, 1996; Nobles et al., 2001). Cellulose is a component of the plant cell wall, which is a complicated mixture of polysaccharides, including also heteroxylan, $(1\rightarrow3)(1\rightarrow4)$-ß-D-glucan, pectin, lignin, and other polymers.

Properties and structure

Unlike most other polymers, cellulose is a homopolymer composed of a group of linear chains of β-$(1\rightarrow4)$ linked glucose units. The molecular weight of the glucan chains must be at least 30–40 kDa to be considered as cellulose. Depending on the source of this polymer, its physical properties, such as crystalline state, degree of crystallization, and molecular weight, can be highly variable. The crystalline state of the cellulose is determined by the arrangement of the glucan chains. In nature, most cellulose (belonging to more than 99% of organisms) is largely produced as crystalline cellulose with some amorphous regions. In crystalline cellulose, the glucan chains are organized in a specific way with respect to one another, whereas no specific pattern is seen in noncrystalline or amorphous cellulose.

Methods for Analysis of Carbohydrate Metabolism in Photosynthetic Organisms.
DOI: http://dx.doi.org/10.1016/B978-0-12-803396-8.00013-2

The crystalline form is generally categorized as cellulose I. The glucan chains in this type of biopolymer are parallel to one another and packaged in the form of submicroscopic rods known as microfibrils, which often can be tens of micrometers in length (Guerriero et al., 2010). The shape and diameter of cellulose microfibrils varies from species to species (Kumar and Turner, 2015).

Different quantities of cellulose I suballomorphs (called I-α and I-β) are generally obtained from natural sources (Atalla and Vanderhart, 1984). Cellulose I-α and I-β differ in their crystalline packaging, molecular structure, and hydrogen bonds. Such are these differences that they could influence the physical properties of the cellulose (Nishiyama et al., 2003). The cellulose of some algae and bacteria can be rich in type I-α, while the cellulose of cotton, wood, and certain tunicates is rich in type I-β (Sugiyama et al., 1991). On the other hand, cellulose II, synthesized by only a few organisms, has antiparallel glucan chains and an intersheet H-bonding that confers the greatest thermodynamic stability (Brown, 1996).

The identification and characterization of cellulose can be achieve by: (1) complete hydrolysis to only glucose; (2) degradation by cellulases; (3) insolubility in NaOH; (4) confirmation of (1→4) linkage after methylation analysis; (5) microfibrils and rods observation by electron microscopy; (6) verification by different analyses, such as NMR, Raman, IR spectroscopy, X-ray, electron diffraction (Brown, 1996).

13.2 Cellulose Biosynthesis and Degradation

Cellulose biosynthesis is a basic biochemical process in land plants. It is essential for cell growth and division. Many hypothetical models have been proposed over the years to explain cellulose biogenesis and crystallization in plants (Delmer, 1999; Brown and Saxena, 2000; Doblin et al., 2002; Jarvis, 2013; Newman and Hill, 2013; Kumar and Turner, 2015). Cellulose synthesis is under a highly controlled enzymatic super complex, located at the tip of the microfibril. The high degree of organization is observed as highly ordered membrane-associated structures. This multimeric complex is a hexameric structure in the form of a rosette within the plasma membrane (Kudlicka and Brown, 1997). Each of the six components of a rosette is expected to synthesize four to six of the glucan chains. Then, 24–36 chains are assembled into a functional microfibril (Doblin et al., 2002). The rosette (of approximately 20–30 nm in diameter) has been immunologically demonstrated to contain cellulose synthase proteins (Kimura et al., 1999; Li et al., 2013). Rosettes were described from several different species (Kumar and Turner, 2015).

Both the isolation and purification of the intact cellulose complex from plants, and the achievement of in vitro cellulose biosynthesis have been extremely difficult. One of the problems is that protein extracts synthesized both cellulose (β-(1→4)-glucans) and callose (β-(1→3)-glucans) and the two activities are hard to separate.

In plants, cellulose synthase catalyzes the polymerization of β-(1→4) glucan chains, where UDP-glucose is the main glucosyl donor in the reaction (Somerville, 2006). This nucleotide sugar is synthesized by the enzyme UDP-glucose pyrophosphorylase from UTP and glucose-1-phosphate (see Chapter 15: Case Study: Nucleotide Sugars). The transfer of glucose from UDP-glucose to cellulose was first described using the particulate fraction of a cell-free extract of *Acetobacter xylinum* and subsequent studies confirmed the role of UDP-glucose as the main precursor in plant cellulose synthesis in developing cotton fibers and in several plants and algae (Carpita and Delmer, 1981; Lin and Brown, 1989; Okuda et al., 1993; Kudlicka and Brown, 1997; Him et al., 2002). The enzyme sucrose synthase (which can also produce UDP-glucose but from sucrose and UDP) has been suggested to be directly involved in cellulose biosynthesis by generating an UDP-glucose channel to the catalytic subunits of cellulose synthase (Amor et al., 1995). Although this source of UDP-glucose raised controversy (Barratt et al., 2009), it remains likely sucrose synthase contributes to providing UDP-glucose in some extent to glucan synthesis (Baroja-Fernandez et al., 2012).

Put simply, cellulose synthesis reaction is a one-step process of polymerization involving the transfer of glucosyl groups by inversion of the configuration of the anomeric carbon. In this type of reaction, a single molecule of cellulose synthase is capable of initiating, elongating, and terminating a β-(1→4) glucan chain. This mechanism implies that the enzyme cellulose synthase binds directly to the substrate (UDP-glucose) and that it is capable of initiating the synthesis of the glucan chain without a primer. Furthermore, cellulose synthase is a highly processive enzyme and remains bound to the growing glucan chain without needing to pull away and rebind during the synthesis process.

The state of knowledge of the plant cell wall biosynthesis, involving cellulose and other (1→4)-β-D-glycans synthesis has been recently reviewed (Carpita, 2012). Despite important recent advances using different approaches (molecular genetics, image analyses, spectroscopic tools, among others), many aspects of cellulose synthesis remain still unknown (Li et al., 2014; Kumar and Turner, 2015). On the other hand, changes in the cell wall can be caused by cellulose breakdown by cell wall hydrolases, including those encoded by a large family of endo-(1→4)-β-D-glucanase (EC 3.2.1.4) genes (referred to as cellulases). However, little is known about the role of these enzymes in plant development and cell wall changes (Del Campillo, 1999). Total cellulose degradation can be achieved by at least three enzyme types, which can be found in other organisms. These enzymes differ in their mode of action and in the substrate used: (1) endo-(1→4)-β-D-glucanases are responsible for splitting the macromolecular chains of cellulose, generating shorter oligomeric units with lower crystallinity, while at the same time generating new free ends for subsequent enzymatic action; (2) exo-(1→4)-β-D-glucanases (cellobiohydrolases) which attack the nonreducing end of the shorter chains producing cellobiose disaccharide molecules as products of the reaction in a synergistic way with the endoglucanase; and (3) β-(1→4)-glucosidases, enzymes responsible for hydrolyzing cellobiose to glucose molecules.

Experimental Procedures

13.3 Cellulose Extraction and Determination

The methods used for cellulose extraction vary according to the required precision or the utilization of the data (eg, for cellulose structure, for content determination, or industrial production) (Pappas et al., 2002; Foston et al., 2011; Newman and Hill, 2013; Zhao et al., 2015). Most general isolation procedures result in a crude cellulose preparation, which is the insoluble residue in a strong sodium hydroxide solution that contains α-cellulose. On the other hand, for lignified cell walls, a delignification step is necessary prior to cellulose isolation. This step, which yields the so-called holocellulose, consists in extracting and removing lignin with chlorination, including extractions with hot alcoholic solutions of organic bases, treatment with an acidic sodium chlorite solution or extraction with diluted peracetic acid (Franz and Blaschek, 1990). This procedure has numerous modifications.

Two general protocols are described below: (1) a method for the fractionation of plant cell wall to obtain pure cellulose (Section 13.3.1) and (2) a procedure for cellulose determination without extraction, used for plants, algae, and cyanobacteria (Section 13.3.2).

13.3.1 Cellulose Extraction After Cell Wall Fractionation

Principle

A cell wall fractionation procedure (originally described by Carpita (1984) for maize cell walls) is used for *Arabidopsis* tissues (Heim et al., 1991). This procedure with some modifications is employed for *Arabidopsis* root cell walls (Peng et al., 2000), and for cotton fibers and seedlings (Li et al., 2013). Basically, the fractionation consists of extracting plant material with phosphate buffer to obtain neutral and acidic polymers in the supernatant. In successive steps, the pellet is treated with chloroform/methanol (to remove lipids), dimethyl sulfoxide (to extract starch), and ammonium oxalate (to remove pectins). The remaining pellet is further treated with KOH containing sodium borohydride and the neutralized supernatant contains hemicelluloses. The KOH nonextractable residue is defined as crude cellulose.

Reagents

Potassium phosphate buffer (pH 7.0)	0.5 M
Chloroform:methanol mixture	1:1 (v/v)
Methanol	
Dimethyl sulfoxide (DMSO):water	9:1 (v/v)
Ammonium oxalate	0.5 %
KOH	0.1 M, containing 1 $mg \cdot mL^{-1}$ of sodium borohydride ($NaBH_4$)

(Continued)

(Continued)

KOH	4 M, containing 1 $mg \cdot mL^{-1}$ $NaBH_4$
Acetic acid:nitric acid:water mixture	8:1:2 (v/v/v)
Trifluoroacetic acid (TFA)	2 M

Procedure

Freeze-dry the plant tissue sample (c.100–150 *Arabidopsis* roots or the equivalent in weight of seedlings). Add 3 mL of cold 0.5 M potassium phosphate buffer (pH 7.0) and grind in a mortar and pestle. Transfer the homogenate to a centrifuge tube, adding 2 mL of the same buffer twice. Centrifuge at 2100 × g for 15 min (supernatant contains neutral and acid polymers such as pectins). Wash the pellet (crude cell wall fraction), twice with 2 mL of the same buffer, and twice with 2 mL of distilled water. Add under stirring and in turn: (1) 3 mL of chloroform:methanol (1:1, v/v) at 40°C for 1 h, and repeat the procedure; (2) 2 mL of methanol at 40°C for 30 min; and (3) 3 mL of distilled water twice. In the case of seedling extraction, repeat the whole procedure. Extract the pellet successively with: (1) DMSO:water (9:1, v/v) overnight under nitrogen, wash twice with 3 mL of DMSO: water, and three times with 3 mL of distilled water; (2) 3 mL of 0.5% ammonium oxalate, at 100°C for 1 h, and wash three times with 3 mL of distilled water; (3) 3 mL of 0.1 M of KOH containing 1 $mg \cdot mL^{-1}$ of sodium borohydride at 25°C for 1 h under a nitrogen atmosphere (this step is repeated once for roots or twice for whole seedlings). Finally, wash with 2 mL of water; (4) 3 mL of 4 M of KOH containing 1 $mg \cdot mL^{-1}$ sodium borohydride, at 25°C for 1 h under a nitrogen atmosphere (this step is repeated once for roots or twice for whole seedlings).

Add 3 mL of acetic acid:nitric acid:water (8:1:2, v/v/v), boil with intermittent stirring for 1 h (as described by Updegraff, 1969). Wash the pellet twice with water. Finally, digest the pellet with 2 M trifluoroacetic acid in a sealed tube at 120°C in an autoclave. The final pellet is crystalline cellulose which is quantified as glucose using the anthrone–sulfuric acid method (see Chapter 1: Determination of Carbohydrates Metabolism Molecules).

13.3.2 Cellulose Determination Without Extraction

Principle

The cell-wall samples are boiled in acetic-nitric reagent. Under this condition, all other polysaccharides (such as lignin, hemicellulose, and xylans) are degraded and removed by centrifugation, and α-cellulose from the insoluble residue is hydrolyzed to sugars and quantified (Updegraff, 1969; York et al., 1986).

Reagents

Acetic acid:nitric acid:water mixture	8:1:2 (v/v/v)

Procedure

Weigh the plant tissue or cells (c.45–55 mg) in eppendorf tubes with screw caps. Add 1.5 mL of the acetic-nitric mixture (8:1:2, v/v/v) to each tube and shake with a vortex. Heat the tubes in a 100°C water bath for 1 h. Allow the tubes to cool at room temperature. Centrifuge the tubes in a swing-out rotor at 11,700 × g for 5 min. Remove the supernatant and wash the precipitate three times with distilled water. For each washing, agitate the tubes and centrifuge at 11,700 × g for 5 min. Dry the tubes in a speed-vac at high heat for 3 h (or low heat overnight). Weigh the tubes after complete drying. Quantitation of total glucose is performed using the anthrone–sulfuric acid method (see Chapter 1: Determination of Carbohydrates Metabolism Molecules).

Comments

Let the tubes dry completely. Avoid adding ethanol, which could interfere with the cellulose determination method. This procedure was used for cellulose determination from cyanobacterial cell walls (Zhao et al., 2015).

13.4 Cellulose and Callose Synthesis Assays

Principle

Plasma membranes from higher plant cells contain processive glycosyltransferases that catalyze the synthesis of both (1→4)-β-D-glucan (cellulose) and (1→3)-β-D-glucan (callose), from UDP-glucose. These glucan synthases are prepared from microsomal membranes treated with detergents from *Arabidopsis* cells (Axelos et al., 1992). The resulting enzyme preparation is incubated in the presence of radioactive UDP-glucose and cellobiose. The labeled products correspond to both glucan synthases or can be discriminated after degradation with enzymes that specifically hydrolyze cellulose or (1→3)-β-D-glucans (Him et al., 2002).

Reagents

Extraction from microsomal membranes

Extraction buffer	
3-(N-morpholino)propanesulfonic acid (MOPS)/NaOH buffer (pH 7.0)	100 mM, containing 2 mM EDTA and 2 mM EGTA
Glycerol	
Taurocholate	
Brij 58	
Glucan synthase activity assay	
MOPS-NaOH buffer (pH 6.8)	0.5 M

(Continued)

(Continued)

Cellobiose	400 mM
$CaCl_2$	80 mM
UDP-glucose	20 mM
UDP-D-[U-^{14}C]-glucose	2.9 μCi
Ethanol	66% and 96%
Acetate buffer (pH 6.5)	0.1 M

Procedure

Enzyme extraction from microsomal membranes

Disrupt the cells (20–25 g) resuspended in 100 mM MOPS-NaOH buffer (pH 7.0) containing 2 mM EDTA and 2 mM EGTA (1 g of cells in 1 mL buffer), using a French Press (at a pressure of 80 megapascals). Centrifuge the homogenate at 5000 × g for 10 min and filter the supernatant through two layers of Miracloth. Centrifuge the filtrate at 150,000 × g for 1 h. Resuspend the pellet (microsomal membranes) in 2 mL extraction buffer containing 10% (v/v) glycerol (Him et al., 2001).

To extract the proteins bound to membranes, incubate the homogenate for 30 min at 4°C, under continuous stirring, in the presence of detergents at their critical micellar concentrations (0.3% taurocholate or 0.05% Brij 58). Centrifuge the microsomal membrane preparation at 150,000 × g for 1 h. The supernatant contains cellulose synthase and (1→3)-β-D-glucan synthase activities, which are stable for several hours at 4°C.

Glucan synthase assay

In a 200-μL total volume, add 20 μL of 0.5 M MOPS-NaOH buffer (pH 6.8), 20 μL of 400 mM cellobiose, 16 μL of $CaCl_2$, 20 μL of 20 mM UDP-glucose, 0.04–0.05 μCi of UDP-D-[U-^{14}C]-glucose, and 50 μL of enzyme preparation. Incubate at 25°C for 2 h, and stop the reaction by addition of 400 μL of ethanol. Precipitate at −20°C for 2 h and filtrate on a glass-fiber filter. Wash the filters successively with 4 mL of 66% ethanol, 4 mL of water, and 4 mL of ethanol. Dry filters, add 5 mL of liquid cocktail, and count radioactivity using a scintillation counter.

To determine the yield of cellulose synthesis, stop the reaction described above by centrifugation. Wash the radioactive products with distilled water. Resuspend in 200 μL of 0.1 M acetate buffer (pH 6.5). Incubate the radioactive products with enzymes that specifically hydrolyze cellulose or (1→3)-β-D-glucan as described by Him et al. (2002). Recover the residual radioactive products from the different incubations by filtration on glass-fiber filters (Millipore Corp.). Wash the filters successively with 4 mL of 66% ethanol, 4 mL of water, and 4 mL of 96% ethanol and dry them. Measure the radioactivity in 5 mL of liquid scintillation mixture, using a scintillation counter.

Comments

While the reactions with the taurocholate-extracted enzyme preparation do not require any cation, the reactions with Brij 58-extracted enzyme preparation require the addition of 8 mM Mg_2SO_4 (final concentration) in the incubation mixture.

The hydrolytic enzymes proposed by Him et al. (2002) are cellobiase (Novozyme 188, 40 $mg \cdot mL^{-1}$), cellobiohydrolase I (CBHI or Cel7A, 40 $mg \cdot mL^{-1}$), cellobiohydrolase II (CBHII or Cel6A, 40 $mg \cdot mL^{-1}$), endoglucanase V (EG V or Cel45, 10 $mg \cdot mL^{-1}$), and $(1 \rightarrow 3)$-β-D-endoglucanase from barley (0.5 $mg \cdot mL^{-1}$) that specifically hydrolyzes $(1 \rightarrow 3)$-β-D-glucans. Cel7A, Cel6A, and Cel45 are recombinant enzymes from *Humicola insolens* expressed in *Aspergillus oryzae* (Schülein, 1997, Novo Nordisk, Denmark).

13.5 Cellulose Degradation

Cellulose can be degraded by: (1) enzymatic hydrolysis or (2) acid hydrolysis. In the first case, a total cellulase system refers to a group of enzymes (endoglucanases, exoglucanases, and β-D-glucosidases) that contributes to the degradation of cellulose to its monomer glucose (Nordmark et al., 2007; Zhang et al., 2009).

13.5.1 Total Cellulase Activity Assay

Principle

Total cellulase activities are measured using insoluble substrates (such as Whatman N°1 filter paper, cotton linter, microcrystalline cellulose, algal cellulose, cellulose containing substrates, α-cellulose, pretreated lignocellulose). The most common assays are: (1) the filter paper assay (FPA), recommended by the International Union of Pure and Applied Chemistry (IUPAC) and (2) the anaerobic cellulase assay (Nordmark et al., 2007).

Filter paper assay

This assay is based on a fixed degree of conversion of substrate, ie, a fixed amount (2 mg) of glucose (based on reducing sugars measured by a colorimetric assay) released from 50 mg of filter paper within a fixed time (1 h). The strengths of this assay are that the substrate is widely available and it is reasonably susceptible to cellulase activity.

Reagents

Paper Whatman N°1
Cellulase preparation
Sodium citrate buffer (pH 4.8) 50 mM
NaOH solution

Procedure

Place a rolled filter paper strip (50 mg, 1.0 × 6.0 cm) into each test tube (13 × 100 cm). Add 1.0 mL of 50 mM citrate buffer (pH 4.8) making sure that the strip is covered. Equilibrate tubes with buffer and substrate to 50°C. Add 0.5 mL of enzyme preparation diluted in citrate buffer. Make at least two dilutions: one dilution must release slightly more than 2.0 mg of glucose and the other slightly less than 2.0 mg of glucose. Depending on the enzyme, this target may be difficult to achieve and additional dilutions must be run. Incubate at 50°C for 60 min. Remove each tube from the 50°C bath and stop the enzyme reaction by immediately adding NaOH to pH 7. The glucose produced is determined by any of the methods described in Chapter 1, Determination of Carbohydrates Metabolism Molecules. Calculate the real glucose concentrations released according to a standard sugar curve.

Comments

Three reaction controls must be simultaneously incubated: (1) reagent blank (1.5 mL of citrate buffer, pH 4.8); (2) enzyme control (1.0 mL of citrate buffer, pH 4.8, plus 0.5 mL enzyme dilution); and (3) substrate control (1.5 mL of citrate buffer, pH 4.8, plus filter paper strip).

Total cellulase activity is described in terms of "filter-paper units" (FPU) per milliliter of original (undiluted) enzyme solution. The FPA has long been recognized for its susceptibility to operator error.

13.5.2 Cellulose Acid Hydrolysis

Principle

This method is based on the complete hydrolysis of cellulose into glucose by measuring the action of diluted acid and subsequent heating to 100–120°C.

Reagents

Cellulose (filter paper, cotton fibers, extracted cellulose)	
Sulfuric acid	3–6% (v/v)
KOH solution	

Procedure

Add 0.2 g of cellulose to 10 mL of 5% sulfuric acid, in a slightly capped tube. Heat the tube containing the reaction mixture for at least 2 h at 90°C. Take 1-mL samples at regular intervals to determine when total hydrolysis has been reached. Stop the hydrolysis by neutralization and slight reversal of the pH by adding potassium hydroxide. Quantify the amount of released glucose by one of the methods described in Chapter 1, Determination of Carbohydrates Metabolism Molecules.

Comments

Alternatively, 5% hydrochloric acid can be used instead of sulfuric acid to achieve cellulose hydrolysis, following the same procedure previously described.

Further Reading and References

Amor, Y., Haigler, C.H., Johnson, S., Wainscott, M., Delmer, D.P., 1995. A membrane-associated form of sucrose synthase and its potential role in synthesis of cellulose and callose in plants. Proc. Natl. Acad. Sci. USA 92, 9353–9357.

Atalla, R.H., Vanderhart, D.L., 1984. Native cellulose: a composite of two distinct crystalline forms. Science. 223, 283–285.

Axelos, M., Curie, C., Mazzolini, L., Bardet, C., Lescure, B., 1992. A protocol for transient gene expression in *Arabidopsis thaliana* protoplasts isolated from cell suspension cultures. Plant Physiol. Biochem. 30, 123–128.

Baroja-Fernandez, E., Munoz, F.J., Li, J., Bahaji, A., Almagro, G., Montero, M., et al., 2012. Sucrose synthase activity in the *sus1/sus2/sus3/sus4 Arabidopsis* mutant is sufficient to support normal cellulose and starch production. Proc. Natl. Acad. Sci. USA 109, 321–326.

Barratt, D.H.P., Derbyshire, P., Findlay, K., Pike, M., Wellner, N., Lunn, J., et al., 2009. Normal growth of *Arabidopsis* requires cytosolic invertase but not sucrose synthase. Proc. Natl. Acad. Sci. USA 106, 13124–13129.

Brown Jr., R.M., 1996. The biosynthesis of cellulose. J. Macromol. Sci. 10, 1345–1373.

Brown Jr., R.M., Saxena, I.M., 2000. Cellulose biosynthesis: a model for understanding the assembly of biopolymers. Plant Physiol. Biochem. 38, 57–67.

Carpita, M.C., Delmer, D.P., 1981. Concentration and metabolic turnover of UDP-glucose in developing cotton fibers. J. Biol. Chem. 256, 308–315.

Carpita, N.C., 1984. Fractionation of hemicelluloses from maize cell walls with increasing concentrations of alkali. Phytochemistry. 23, 1089–1093.

Carpita, N.C., 2012. Progress in the biological synthesis of the plant cell wall: new ideas for improving biomass for bioenergy. Curr. Opin. Biotechnol. 23, 330–337.

Del Campillo, E., 1999. Multiple endo-1, 4-β-D-glucanase (cellulase) genes in *Arabidopsis*. Curr. Top. Dev. Biol. 46, 39–61.

Delmer, D.P., 1999. Cellulose biosynthesis: exciting times for a difficult field of study. Annu. Rev. Plant Physiol. Plant Mol. Biol. 50, 245–276.

Doblin, M.S., Kurek, I., Jacob-Wilk, D., Delmer, D.P., 2002. Cellulose biosynthesis in plants: from genes to rosettes. Plant Cell Physiol. 43, 1407–1420.

Foston, M.B., Hubbell, C.A., Ragauskas, A.J., 2011. Cellulose isolation methodology for NMR analysis of cellulose ultrastructure. Materials. 4, 1985–2002.

Franz, G., Blaschek, W., 1990. Cellulose. In: Dey, P.M., Harborne, J.B. (Eds.), Methods in Plant Biochemistry, Vol. 2 Carbohydrates. Academic Press, London, pp. 291–322.

Guerriero, G., Fugelstad, J., Bulone, V., 2010. What do we really know about cellulose biosynthesis in higher plants? J. Integ. Plant. Biol. 52, 161–175.

Heim, D.R., Skomp, J.R., Waldron, C., Larrinua, I.M., 1991. Differential response to isoxaben of cellulose biosynthesis by wild-type strains of *Arabidopsis thaliana*. Pest. Biochem. Physiol. 39, 93–99.

Him, L.K.J., Pelosi, L., Chanzy, H., Putaux, J.-L., Bulone, V., 2001. Biosynthesis of (1→3)-β-D-glucan (callose) by detergent extracts of a microsomal fraction from *Arabidopsis thaliana*. Eur. J. Biochem. 268, 4628–4638.

Him, L.K.J., Chanzy, H., Müller, M., Putaux, J.-L., 2002. In vitro versus in vivo cellulose microfibrils from plant primary wall synthases: structural differences. J. Biol. Chem. 27, 36931–36939.

Jarvis, M.C., 2013. Cellulose biosynthesis: counting the chains. Plant Physiol. 163, 1485–1486.

Kimura, S., Laosinchai, W., Itoh, T., Cui, X., Linder, C.R., Brown Jr., R.M., 1999. Immunogold labeling of rosette terminal cellulose synthesizing complexes in the vascular plant *Vigna angularis*. Plant Cell. 11, 2075–2086.

Kudlicka, K., Brown Jr., R.M., 1997. Cellulose and callose biosynthesis in higher plants. Solubilization and separation of (1→3)- and (1→4)-beta-glucan synthase activities from mung bean. Plant Physiol. 115, 643–656.

Kumar, M., Turner, S., 2015. Plant cellulose synthesis: CESA proteins crossing kingdoms. Phytochemistry. 112, 91–99.

Li, A., Xia, T., Xu, W., Chen, T., Li, X., Fan, J., et al., 2013. An integrative analysis of four CESA isoforms specific for fiber cellulose production between *Gossypium hirsutum* and *Gossypium barbadense*. Planta. 237, 1585–1597.

Li, S., Bashline, L., Lei, L., Gu, Y., 2014. Cellulose synthesis and its regulation. Arabidopsis Book. 12, e0169, Published online 2014 January 13. htttp://dx.doi.org/10.1199/tab.0169.

Lin, F.C., Brown Jr., R.M., 1989. Purification of cellulose synthase from *Acetobacter xylinum*. In: Schuerch, C. (Ed.), Cellulose and Wood: Chemistry and Technology. Wiley Intersci., New York, pp. 473–492.

Newman, R.H., Hill, S.J., 2013. Wide-angle X-ray scattering and solid state nuclear magnetic resonance data combined to test models for cellulose microfibrils in mung bean cell wall. Plant Physiol. 163, 1558–1567.

Nishiyama, Y., Sugiyama, J., Chanzy, H., Langan, P., 2003. Crystal structure and hydrogen bonding system in cellulose I(alpha) from synchrotron X-ray and neutron fiber diffraction. J. Am. Chem. Soc. 125, 14300–14306.

Nobles, D.R., Romanovicz, D.K., Brown Jr., R.M., 2001. Cellulose in cyanobacteria. Origin of vascular plant cellulose synthase? Plant Physiol. 127, 529–542.

Nordmark, T.S., Bakalinsky, A., Penner, M.H., 2007. Measuring cellulase activity: application of the filter paper assay to low-activity enzyme preparations. Appl. Biochem. Biotechnol. Humana Press.

Okuda, K., Li, L., Kudlicka, K., Kuga, S., Brown Jr., M.R., 1993. β-Glucan synthesis in the cotton fiber. I. Identification of β-1,4- and β-1,3-glucans synthesized in vitro. Plant Physiol. 101, 1131–1142.

Pappas, C., Tarantilis, P.A., Daliani, I., Mavromoustakos, T., Polissiou, M., 2002. Comparison of classical and ultrasound-assisted isolation procedures of cellulose from kenaf (*Hibiscus cannabinus* L.) and eucalyptus (*Eucalyptus rodustrus* Sm.). Ultrason. Sonochem. 9, 19–23.

Peng, L.C., Hocart, C.H., Redmond, J.W., Williamson, R.E., 2000. Fractionation of carbohydrates in *Arabidopsis* root cell walls shows that three radial swelling loci are specifically involved in cellulose production. Planta. 211, 406–414.

Schülein, M., 1997. Enzymatic properties of cellulases from *Humicola insolens*. J. Biotechnol. 57, 71–81.

Somerville, C., 2006. Cellulose synthesis in higher plants. Annu. Rev. Cell Dev. Biol. 22, 53–78.

Sugiyama, J., Vuong, R., Chanzy, H., 1991. Electron diffraction study on the two crystalline phases occurring in native cellulose from an algal cell wall. Macromol. 24, 4168–4175.

Updegraff, D.M., 1969. Semimicro determination of cellulose in biological materials. Anal. Biochem. 32, 420–424.

York, W.S., Darvill, A.G., McNeil, M., Stevenson, T.T., Albersheim, P., 1986. Isolation and characterization of plant cell walls and cell wall components. Methods Enzymol. 118, 3–40.

Zhang, Y.H.P., Hong, J., Ye, X., 2009. Cellulase assays. In: Mielenz, J.R. (Ed.), Biofuels: Methods and Protocols, Methods in Molecular Biology, vol. 581. Humana Press, pp. 213–231.

Zhao, C., Li, Z., Li, T., Zhang, Y., Bryant, D.A., Zhao, J., 2015. High-yield production of extracellular type-I cellulose by the cyanobacterium *Synechococcus* sp. PCC 7002. Cell Discov. 1, 15004.

CHAPTER 14

Case Study: Sugar Phosphates

Chapter Outline

14.1 Introduction

Phosphorus is an essential nutrient for living beings. To plants, it is the second element in importance, which is taken up from soil (as inorganic phosphate) through their radical system, and incorporated to different phosphoric compounds, such as sugar phosphates, nucleotide phosphates, phospholipids, or phosphoproteins. Interestingly, phosphorus in animals originates from plant phosphate compounds.

Sugar phosphates are defined as carbohydrates to which a phosphate group is bound by an ester or an ether linkage, depending on whether it involves an alcoholic or a hemiacetalic hydroxyl, respectively. The analysis of both types of sugar phosphates requires the knowledge of their physical and chemical properties (eg, solubility, acid hydrolysis rates, acid strengths, and ability to act as sugar group donors).

Sugar phosphates are closely associated with the photosynthetic carbon reduction cycle, and are key molecules in metabolism, being important intermediates in glycolysis, gluconeogenesis,

Methods for Analysis of Carbohydrate Metabolism in Photosynthetic Organisms.
DOI: http://dx.doi.org/10.1016/B978-0-12-803396-8.00014-4

and oxidative pentose phosphate pathways. They are also involved in the synthesis of other phosphate compounds, in metabolic regulation, and signaling. The properties of sugar phosphates and the knowledge of reactions involved in their formation or cleavage, crucial in the biochemistry of photosynthetic organisms, are described in this chapter.

14.2 General Properties of Sugar Phosphates

14.2.1 Acid and Alkaline Hydrolysis

One of the properties that contribute to the chemical identification of phosphoric esters is the measurement of the hydrolysis rate, which is affected by several factors (like temperature, pH, the carbon chain structure and the phosphate group position, and presence of other substituents).

Phosphates linked to a hemiacetalic hydroxyl group are more acid-labile than those esterifying an alcoholic group. Consequently, in general, aldoses-1-phosphate (such as mannose-1-phosphate and glucose-1-phosphate) hydrolyze in a few minutes in 0.1 N acid solution at 100°C, while aldose 2-deoxyribose-1-phosphate and the ketose fructofuranose-2-phosphate (the most acid-labile sugar phosphate described so far) are hydrolyzed at pH 4 at room temperature. On the contrary, the hydrolysis of aldoses-6-phosphate (such as glucose-6-phosphate, fructose-6-phosphate, mannose-6-phosphate, and galactose-6-phosphate) requires more than 100 h in 1 N acid solution at 33°C. Other compounds with a phosphate group attached to an alcoholic hydroxyl group (eg, glycerophosphate and glyceric acid phosphate) are very stable to acid hydrolysis.

On the other hand, the anomeric isomers are differentially hydrolyzed in 1 N hydrochloric acid at 33°C (β anomers are generally more labile than α anomers). For example, the hydrolysis constant values (k) for α-glucose-1-phosphate and β-glucose-1-phosphate are 7.8×10^{-3} min^{-1} and 16.4×10^{-3} min^{-1}, respectively. The hydrolysis rate of the phosphate at C-1 is also influenced by a substitution in C-6 (eg, a phosphate group at this position, like in glucose-1,6-diphosphate, decreases the rate of hydrolysis). A similar hydrolysis effect is found in glucuronic-1-phosphate acid whose k becomes 10–20 times smaller than that of glucose-1-phosphate. Conversely, a decrease in size of the groups located at C-6 increases the hydrolysis rate (Pontis and Leloir, 1972).

Hydrolysis rate of aldoses-1-phosphate is also affected by substituents at C-2. For example, the acid lability of 2-deoxyribose-1-phosphate is higher than that of ribose-1-phosphate, and 2-deoxyamino sugars are more stable than the corresponding nonsubstituted sugars (Brown, 1953; Cardini and Leloir, 1953). However, α-glucosamine-1-phosphate exhibits similar hydrolysis behavior to α-D-glucose-1-phosphate at 100°C and at pH range between 3 and 6 (Ramsay and Pizanis, 1965).

Regarding ketoses, the phosphate acid hydrolysis rate depends on the distance to the carbonyl group (eg, k value for fructose-1-phosphate is approximately 16 times higher than that for fructose-6-phosphate). In turn, D-fructofuranose-2-phosphate (a kestose-2-phosphate) is more acid-labile than glucose-1-phosphate (the corresponding aldose-1-phosphate) (Pontis and Fischer, 1963).

The carbohydrate structure (pyranose or furanose form) also is a determining factor on the acid lability. For example, fructofuranose-2-phosphate is more labile than fructopyranose-2-phosphate (Pontis and Fischer, 1963); fructose-6-phosphate, which can only form a furanose ring, is more labile than glucose-6-phosphate; and ribose-5-phosphate, which can form a furanose ring, is more stable than ribulose-5-phosphate that only exists in a linear form. Also, the lack of ring structures seems to contribute to the bound acid lability (such is the case of glycolaldehyde phosphate, erythrose-4-phosphate, and erythrulose-phosphate) (Pontis and Leloir, 1972).

The presence of a free aldehyde or keto group in a sugar phosphate leads to a rapid alkaline degradation. On the contrary, aldoses-1-phosphate and kestoses-2-phosphate are stable to alkali (Leloir, 1951; Leloir and Cardini, 1963; Pontis and Fischer, 1963; Degani and Halmann, 1968).

14.2.2 Acid Strength

Sugar phosphates are stronger acids than orthophosphoric acid (apparent sugar phosphate pK values are lower than orthophosphoric acid pK_1 and pK_2, as shown in Table 14.1).

Table 14.1: Apparent ionization constants of some sugar phosphates[a]

Compound	pK_1	pK_2	pK_3
Orthophosphoric acid	1.97	6.82	12
Glyceraldehyde-3-phosphate	2.10	6.75	
Dihydroxyacetone-phosphate	1.77	6.45	
Xylose-1-phosphate	1.25	6.15	
Glucose-1-phosphate	1.10	6.13	
Glucose-3-phosphate	0.84	5.67	
Glucose-4-phosphate	0.84	5.67	
Glucose-6-phosphate	1.54[b]	6.24[b]	
Galactose-1-phosphate	1.00	6.17	
Fructose-6-phosphate	0.97	6.11	
Fructose-1,6-diphosphate	1.48	6.32	
Maltose-1-phosphate	1.52	5.89	
N-acetyl-D-glucosamine 1-phosphate (both α and β)	<1.4	6.0	

[a]Taken partly from Leloir and Cardini (1963).
[b]Measured at 30.5°C.
Source: Republished with permission from Pontis, H.G., Leloir, L.F., 1972. Sugar phosphates and sugar nucleotides. In: Elving, P.J., Kolthoff, I.M. (Eds.), Analytical Chemistry of Phosphorous Compounds. Wiley-Interscience, New York, p. 624.

The acid strength is the important property of these compounds that allows their separation by ion exchange chromatography or electrophoresis (Pontis and Leloir, 1972).

14.2.3 Borate Complex Formation

The formation of acidic complexes between the hydroxyl groups of sugars or sugar phosphates with borate ions has facilitated their separation and analysis by classical methodologies (paper, thin layer, and ion exchange chromatography, and electrophoresis) and instrumental analysis (high-performance anion exchange chromatography, nuclear magnetic resonance spectroscopy, matrix-assisted laser desorption/ionization Fourier transform mass spectrometry) (Pontis and Leloir, 1972; Mopper et al., 1980; Smrcka and Jensen, 1988; Penn et al., 1997; Yamamoto et al., 1999). The structure and configuration of a sugar influence the degree of borate complexation. Borate ions predominantly react with *cis*-α-glycols, giving strong complexes. Hence, sugars with a furanose structure that have vicinal *cis*-hydroxyl groups exhibit a greater degree of borate complexation than those with a pyranose structure.

Experimental Procedures

14.3 Determination of Sugar Phosphates

Many sugar phosphates can be specifically estimated with enzymatic assays. However, colorimetric techniques are generally used either for quantifying the sugar moiety or the phosphate.

14.3.1 Determination of the Phosphate Group

Among the several available colorimetric assays for phosphate determination, the Fiske–Subbarow method is one of the most widely used because its simplicity and specificity. The method is described in Chapter 1, Determination of Carbohydrates Metabolism Molecules.

14.3.2 Determination of the Sugar Moiety

The anthrone method, originally developed for the determination of hexoses, is also a simple and sensitive procedure successfully applied for nonspecific determination of sugar phosphates. Another general method used for this purpose is the estimation of reducing power. Sugar phosphates (aldoses or ketoses) with reducing property arising out of the presence of a free carbonyl group can be determined with the Somogyi–Nelson reagent (see Chapter 1: Determination of Carbohydrates Metabolism

Molecules). For the determination of sugar phosphates such as aldoses-1-phosphate and ketoses-2-phosphate, the phosphate group should be hydrolyzed before determining the sugar moiety as reducing power. Other methods that can be used depend on specific color reactions. For example, fructose-6-phosphate can be determined with the thiobarbituric acid reagent, described in Chapter 1, Determination of Carbohydrates Metabolism Molecules.

14.3.3 Enzymatic Assays

The ability of enzymes to react specifically with a component of a mixture confers them particular value in sugar phosphates analysis (Stitt et al., 1989). The most widely used enzymes for enzymatic analysis are NAD^+- or $NADP^+$-dependent dehydrogenases. Only for a few sugar phosphates is it feasible for the direct use of specific dehydrogenases (eg, α-glycerophosphate dehydrogenase, glyceroaldehyde 3-phosphate dehydrogenase, and glucose 6-phosphate dehydrogenase), which can be used directly to estimate L-α-glycerophosphate, D-glyceraldehyde 3-phosphate, and glucose 6-phosphate, respectively). However, usually it is necessary to couple dehydrogenases with suitable specific enzymes to determine most sugar phosphates. For instance, the coupled enzymes phosphoglucomutase plus glucose 6-phosphate dehydrogenase, phosphoglucose isomerase plus glucose 6-phosphate dehydrogenase, and liver aldolase plus α-glycerophosphate dehydrogenase are used to estimate α-D-glucose 1-phosphate, fructose 6-phosphate, and D-fructose-1-phosphate, respectively (Pontis and Leloir, 1972). Glucose-6-phosphate, fructose-6-phosphate, and mannose-6-phosphate can be determined as glucose-6-phosphate oxidation to phosphogluconate in the presence of $NADP^+$ and glucose-6-phosphate dehydrogenase (see Chapter 1: Determination of Carbohydrates Metabolism Molecules). Other protocols are described below.

14.4 Determination of Fructose-1,6-Diphosphate

Principle

Fructose-1,6-diphosphate analysis is carried out by a coupled assay utilizing fructose 1,6-diphosphate aldolase, triose-phosphate isomerase, and glycerol-3-phosphate dehydrogenase, in the presence of NADH (molar extinction coefficient at 340 nm = 6.220 $M^{-1}cm^{-1}$), whose oxidation to NAD^+ is registered spectrophotometrically.

$$\text{Fructose-1, 6-diphosphate} + H_2O \xrightarrow{\text{Aldolase}} \text{Glyceraldehide-3-phosphate} + \text{Dihydroxyacetone-phosphate}$$

$$\text{Glyceraldehyde-3-phosphate} \xrightarrow{\text{Triose-phosphate isomerase}} \text{Dihydroxyacetone-phosphate}$$

$$2\ \text{Dihydroxyacetone-phosphate} + 2\ \text{NADH} \xleftrightarrow{\text{Glycerol-3-phosphate dehydrogenase}} 2\ \text{Glycerophosphate} + 2\ \text{NAD}^+ + \text{H}^+$$

Reagents

KSH buffer solution:	
Glycil-glycine buffer (neutralized with NaOH) (pH 7.5) containing 200 mM potassium acetate and 50 mM β-mercaptoethanol	100 mM
NADH (sodium salt)	2 mM in 1 mM NaOH
Aldolase (fructose 1,6-diphosphate aldolase)	
α-Glycerophosphate dehydrogenase-triosephosphate-isomerase (mixed crystals, diluted 1:5 in distilled water at 0°C)	$10\ mg \cdot mL^{-1}$
Fructose-1,6-diphosphate (tetra(cyclohexylammonium) salt) (pH 7.5)	20 mM

Notes: KSH buffer solution containing β-mercaptoethanol is stable for 1–2 days. KSH buffer can be replaced by 100 mM Tris-HCl buffer. Store NADH solution at 0–4°C for a period no longer than 2 weeks. Immediately before use, dilute enzymes in cold reaction buffer.

Procedure

In a 1-mL cuvette with 1-cm light path, add 500 μL of KSH buffer solution, 100 μL of sample (or 100 μL of 20 mM fructose-1,6-diphosphate standard solution), 100 μL of NADH, 10 μL of α-glycerophosphate dehydrogenase-triose-phosphate-isomerase enzyme solution and $50\ \mu g \cdot mL^{-1}$ of aldolase. Make up to 1 mL with distilled water. Mix quickly by inversion, place the cuvette in a suitable thermostatted spectrophotometer (at 25°C) and register the absorbance at 340 nm every 10–15 s, or continually until linearity is reached (approximately between 3 to 5 min).

It is assumed that the oxidation of 2 μmol of NADH (12.44 absorbance units) represents the cleavage of 1 μmol of fuctose-1,6-diphosphate under these conditions.

14.5 Determination of Fructose-2,6-Diphosphate

Principle

Fructose-2,6-diphosphate determination is based on the activation of the enzyme fructose-6-phosphate kinase, pyrophosphate-dependent (PPi-PFK, EC 2.7.1.90), which is only active in the presence of fructose-2,6-diphosphate (activator), catalyzing the following reaction:

$$\text{Fructose-6-phosphate} + \text{PPi} \xrightarrow{\text{PPi}-\text{PFK}+\text{fructose}-2,6\text{-diphosphate}} \text{Fructose-1,6-diphosphate} + \text{Pi (inorganic phosphate)}$$

The activation is proportional to the amount of fructose-2,6-diphosphate present. This method is applicable to fructose-2,6-diphosphate determination in homogenates from different tissues (Sabularse and Anderson, 1981; Van Schaftingen et al., 1982).

Reagents

Tris-HCl buffer (pH 8.0)	1 M
$MgCl_2$	200 mM
NADH	3 mM
Aldolase (fructose 1,6-diphosphate aldolase)	
Triose phosphate isomerase	
Glycerol 3-phosphate dehydrogenase	
Fructose-6-phosphate	100 mM
Sodium pyrophosphate (PPi)	20 mM
Fructose-6-phosphate kinase, pyrophosphate-dependent (PPi-PFK) from potato tubers	
Fructose-2,6-diphosphate standard solution	10 μM

Procedure

In a 1-mL total volume, mix 50 μL of Tris-HCl buffer (pH 8.0), 25 μL of $MgCl_2$, 50 μL of NADH, 50 $\mu g \cdot mL^{-1}$ aldolase, 1 $\mu g \cdot mL^{-1}$ triose phosphate isomerase, 10 $\mu g \cdot mL^{-1}$ glycerol 3-phosphate dehydrogenase, 100 μL of fructose-6-phosphate, purified PPi-PFK, and an aliquot of sample. Preincubate for 5 min at 30°C and start the reaction by the addition of 50 μL of PPi. A standard activation curve should be obtained incubating the reaction mixture (omitting the sample) and adding increasing concentrations of fructose-2,6-diphosphate (0–60 nM). Incubate all reactions at 30°C. PPi-PFK activity is measured by the appearance of fructose-1,6-diphosphate which is quantified by following the absorbance at 340 nm due to NADH oxidation (see Section 14.4). The PPi-PFK activation corresponds to the amount of fructose-2,6-diphosphate present, in accordance with the standard curve.

Comments

Fructose-2,6-diphosphate is extracted from plant tissues with 0.1 M glycine/NaOH buffer (pH 10.0) (fresh tissue:buffer, 1:3 w/v). The sample is centrifuged and the supernatant neutralized with acetic acid in the presence of 20 mM Hepes. Fructose-2,6-diphosphate is then quantified by the stimulation of PPi-PFK. In order to confirm that the activation produced by a sugar extract is due to fructose-2,6-diphosphate, it is necessary to heat the sample at pH 4 for 10 min. Under these conditions fructose-2,6-diphosphate is hydrolyzed and the activation effect vanishes. There is no other phosphoric ester that hydrolyzes under this acidic condition.

PPi-PFK from mung beans can be purified as described by Sabularse and Anderson (1981). The enzyme from potato tuber can be purchased and is also stimulated by fructose-2,6-diphosphate (half-maximal activation at 5.5 nM fructose-2,6-diphosphate).

14.6 Determination of Mannose-1-Phosphate and Mannose-6-Phosphate

Principle

Mannose-1-phosphate determination is based on its transformation into mannose-6-phosphate by the enzyme phosphomannose mutase (together with its coenzyme mannose-1,6-diphosphate). The phosphoric ester in C-6 is determined by its transformation to fructose-6-phosphate by the action of phosphomannose isomerase, followed by its conversion into glucose-6-phosphate by phosphoglucose isomerase. The measurement of the last phosphoglucose isomerase reaction product involves the addition of glucose-6-phosphate dehydrogenase (procedure described in Chapter 1: Determination of Carbohydrates Metabolism Molecules, Section 1.5).

$$\text{Mannose-1-phosphate} \xleftrightarrow{\text{Phosphomannosemutase + coenzyme}} \text{Mannose-6-phosphate}$$

$$\text{Mannose-6-phosphate} \xleftrightarrow{\text{Phosphomannose isomerase}} \text{Fructose-6-phosphate}$$

$$\text{Fructose-6-phosphate} \xleftrightarrow{\text{Phosphoglucose isomerase}} \text{Glucose-6-phosphate}$$

$$\text{Glucose-6-phosphate} + \text{NADP}^+ \xrightarrow{\text{Glucose-6-phosphate dehydrogenase}} \text{6-Phosphogluconate} + \text{NADPH} + \text{H}^+$$

Reagents

Tris-HCl buffer (pH 7.5)	0.5 M
α-D-glucose-1,6-diphosphate (added as cofactor)	0.1 mM
$MgCl_2$	100 mM
$NADP^+$	100 mM
Phosphomannose mutase	
Phosphomannose isomerase (from *Escherichia coli*)	$100\ U \cdot mg\ protein^{-1}$
Phosphoglucose isomerase (from yeast)	$400\ U \cdot mg\ protein^{-1}$
Glucose-6-phosphate dehydrogenase (from yeast)	$200\ U \cdot mg\ protein^{-1}$
Mannose-1-phosphate (standard solution)	10 mM

Procedure

In a 100-μL total reaction volume, mix 50 μL of sample (or standard solution containing between 10 and 25 nmol of mannose-1-phosphate), 5 μL of 100 mM $MgCl_2$, 6 μL of 100 mM $NADP^+$, 10 μL of 0.5 M Tris-HCL buffer (pH 7.5), 1 μL of glucose-1,6-diphosphate, 0.3 μg of phosphomannose mutase, 2 μg of phosphomannose isomerase, 1 μg phosphoglucose isomerase, and 1 μg of glucose-6-phosphate dehydrogenase. Incubate at 37°C for 10 min. Make up to 1 mL with distilled water, mix and determine absorbance at 340 nm. To calculate the μmoles of hexose phosphate present in the sample, divide the increase in absorbance by 6.22, taking into account that the absorbance of 1 mL containing 1 μmol of NADPH, in a 1-cm path length cuvette at 340 nm is 6.22.

To determine mannose-6-phosphate, glucose-1,6-diphosphate and phosphomannose mutase are omitted in the reaction mixture. Mannose-6-phosphate determination is carried out as previously described in Chapter 1, Determination of Carbohydrates Metabolism Molecules.

Comments

Phosphomannose mutase from plant origin can be prepared according to Murata (1976) or Small and Matheson (1979). This enzyme requires both Mg^{2+} and cofactor (mannose-1,6-diphosphate or α-D-glucose-1,6-diphosphate) for activity. The cofactor can accompany the phosphomannose mutase preparation. Activity is maximal at pH 6.5–7.0, and the pH used in the proposed protocol is a compromise between the optimal pHs of the auxiliary enzymes.

14.7 Separation of Sugar Phosphates in Ion Exchange Resins

Principle

All sugar phosphates have a negative charge due to the phosphate ion. The sugar phosphate ionization (crucial for the interaction with ion exchange columns) depends only on the phosphate group that takes place at pH between 7 and 8. Under this condition, the separation of different sugar phosphates is not possible. In order to accomplish the separation of sugar phosphates, they must be complexed with borate ions (see Section 14.2.3). The differences in the strengths of the complexes can be used to successfully and rapidly achieve the separation on anion exchangers (in the borate form) with borate solutions as eluents (triethylammonium borate is more stable than ammonium borate solution). The use of ammonium or triethylammonium salts allows an easy (and almost quantitative) recovery of the sugar phosphates by removing the borate after freeze-drying, and methanol addition and evaporation to dryness in a rotary evaporator (this is not possible with the sodium or potassium borate salts). Complexes stability is dependent on the borate ion concentration in the mobile phase and on the pH (pH values between 8 and 9 are suitable for sugar phosphate separation).

The procedure described below to separate several sugar phosphates is based on the use of a linear gradient elution with ammonium or triethyl ammonium borate (Lefebvre et al., 1964).

Reagents

Resin preparation

Dowex-1 chloride form (200–400 mesh) is changed to borate form by passing 0.8 M potassium tetraborate solution until all chloride ions are removed (the effluent should not give precipitate on addition of silver nitrate), and then washed with distilled water.

Preparation of the triethylammoniumtetraborate solution

Triethylammonium (or ammonium) tetraborate solution (pH 7.5) is made by mixing a freshly prepared boric acid solution with triethylamine (or ammonia). For example, to prepare a 0.4 M triethylammonium tetraborate solution, 99.2 g of boric acid and 112 mL of triethylamine are dissolved in distilled water, and made up to 1 L. The sugar phosphate mixture is adjusted to pH 8 with ammonium hydroxide before it is loaded into the column.

Procedure

Column chromatography

Analytical separation of 2–100 μmol of a mixture of sugar phosphates is carried out in a 0.5-cm diameter and 60-cm length Dowex-1 borate column, which is eluted with a linear gradient from 0 to 0.4 M of triethylammonium tetraborate. Fractions (1.4-mL volume) are collected at an elution rate of 1.0–1.5 $mL \cdot min^{-1}$. Eluted sugars are determined by the anthrone reagent method (see Chapter 1: Determination of Carbohydrates Metabolism Molecules). An example of this separation is shown in Fig. 14.1. The fractions under each peak are pooled and freeze-dried or evaporated to dryness in a rotary evaporator. Remaining ammonium borate (or triethylammonium borate) is removed after methanol addition and evaporation to dryness (this procedure is repeated two or three times).

14.8 Separation of Phosphoric Esters by High-Performance Liquid Chromatography

Sugar phosphates in plants and other photosynthetic organisms have been traditionally determined using enzymatic assays which are highly sensitive and very specific techniques (Stitt et al., 1989; Gibon et al., 2002). However, with these methods it is not possible to perform parallel analysis of different compounds in the same sample. To overcome this, different useful alternatives of high-performance liquid chromatography (HPLC) have been developed in the last decades. Additionally, mass spectrometry has also been applied for analyzing sugar phosphates (Antonio et al., 2007).

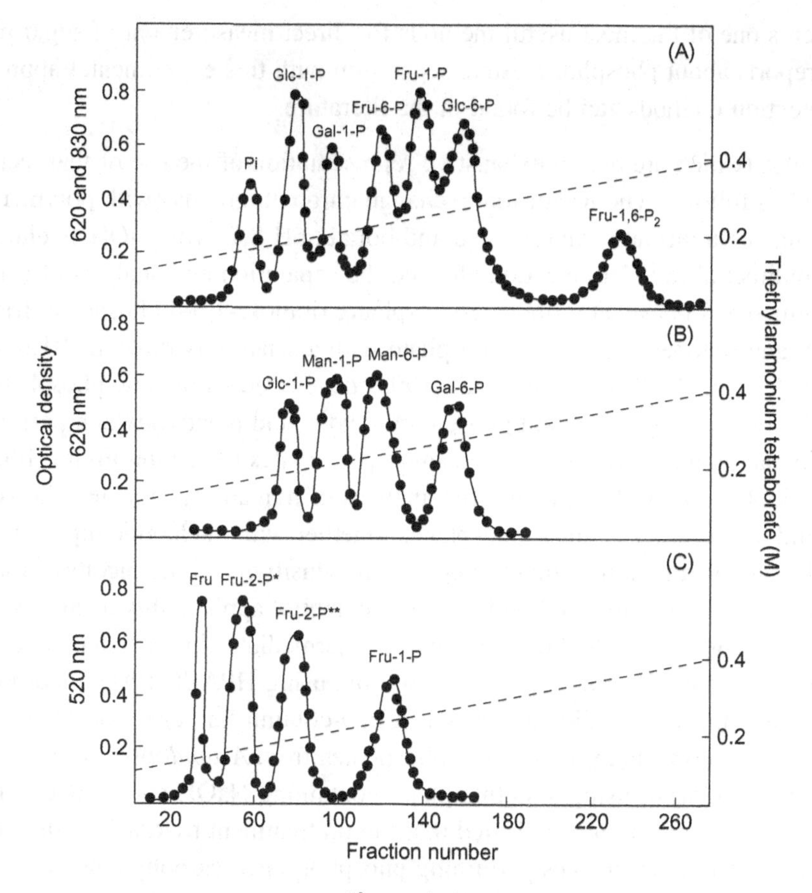

Figure 14.1

Ion exchange chromatography of sugar phosphates in a Dowex-1 borate column. The mixtures applied to the column were: (A) 2 μmol Pi, 9 μmol glucose-1-phosphate (Glc-1-P), 7 μmol galactose-1-phosphate (Gal-1-P), 10 μmol fructose-1-phosphate (Fru-1-P), 10 μmol fructose-6-phosphate (Fru-6-P), 9 μmol glucose-6-phosphate (Glc-6-P), 9 μmol fructose-1,6-diphosphate (Fru-1,6-P_2). (B) 5 μmol Glc-1-P, 18 μmol manose-1-phosphate (Man-1-P), 17 μmol manose-6-phosphate (Man-6-P), 20 μmol galactose-6-phosphate (Gal-6-P). (C) 1 μmol fructose (Fru), 3 μmol fructofuranose-2-phosphate (Fru-2-P*), fructopyranose-2-phosphate (Fru-2-P**), 6 μmol fructose-1-phosphate. The dotted straight line corresponds to the tetraborate and triethylammonium gradient from 0 to 0.4 M. *Reproduced from Lefebvre, M.J., Gonzalez, N., Pontis, H. G., 1964. Anion-exchange chromatography of sugar phosphates with triethylammonium borate. J. Chromatog. 15, 495–500.*

High-performance anion exchange chromatography (HPAEC) is an improved methodology that offers the possibility to simultaneously analyze sugar phosphates and nucleotides involved in the main pathways of photosynthetic organisms. When this technique is coupled with pulsed amperometric detection (PAD), it permits direct quantification of nonderivatized sugars with minimal sample preparations and at a high sensitivity level.

This represents one of the most useful methods for direct measurement of sugar phosphates. Numerous reports about phosphoric esters separation with this experimental approach and different detection methods can be found in the literature.

Some examples (which are not an exhaustive representation of the use of that technique) are described as follows. The use of ion-exchange chromatography (with polymer based anion-exchange columns and trimesic acid and borate, pH 8.7, with LiOH as eluent) coupled to indirect ultraviolet detection allowed the separation and analysis of fructose-6-phosphate, glucose-6-phosphate, ribose-5-phosphate, ribulose-1,5-diphosphate, fructose-1,6-diphosphate, and sedoheptulose-1,7-diphosphate with a sensitivity down to 10 nmol (Smrcka and Jensen, 1988). Ikeguchi et al. (1993) developed an organic phosphate analyzer based on HPAEC and a post column phosphomolybdic acid reaction which permitted the selective separation and detection of some sugar phosphates (detection limit c.100 pmol). A chromatographic approach applied to the analysis of human erythrocytes allowed the resolution and quantitation of sugar phosphates together with nucleotide diphosphate sugars using HPAEC coupled to conductimetric detection (sensitivity in the picomole range) (Hull and Montgomery, 1994). This method has also a potential applicability as an assay system for the analysis of glycolysis and other pathways intermediates in photosynthetic organisms. Another practical method, with improved resolution, using HPAEC-PAD, incorporates a column packed with titanium dioxide resin as a trap-column for sugar phosphates and nucleotides (used for the analysis of sugar phosphates from *Arabidopsis thaliana*) (Sekiguchi et al., 2004) or using a mobile phase containing NaOH and Na_2CO_3 (Jeong et al., 2007). Subsequent studies indicated that the pretreatment by titanium dioxide was prone to plant matrix interferences perturbing phosphorylated carbohydrate detection (Delatte et al., 2009). A final mention to these examples is the development of a sensitive and robust mixed-mode high-performance liquid chromatography–tandem mass spectrometry method used in metabolome analysis, for qualitative and quantitative determination of sugar phosphates in tobacco (Cruz et al., 2008).

Further Reading and References

Antonio, C., Larson, T., Gilday, A., Graham, I., Bergström, E., Thomas-Oates, J., 2007. Quantification of sugars and sugar phosphates in *Arabidopsis thaliana* tissues using porous graphitic carbon liquid chromatography-electrospray ionization mass spectrometry. J. Chromatogr. 1172, 170–178.

Brown, D.H., 1953. Action of phosphoglucomutase on D-glucosamine-6-phosphate. J. Biol. Chem. 204, 877–890.

Cardini, C.E., Leloir, L.F., 1953. Enzymic phosphorylation of galactosamine and galactose. Arch. Biochem. Biophys. 45, 55–64.

Cruz, J.A., Emery, C., Wüst, M., Kramer, D.M., Lange, B.M., 2008. Metabolite profiling of Calvin cycle intermediates by HPLC-MS using mixed-mode stationary phases. Plant J. 55, 1047–1060.

Degani, C., Halmann, M., 1968. Alkaline reactions of D-glucose 6-phosphate. J. Am. Chem. Soc. 90, 1313–1317.

Delatte, T.L., Selman, M.H., Schluepmann, H., Somsen, G.W., Smeekens, S., De Jong, G.J., 2009. Determination of trehalose-6-phosphate in *Arabidopsis* seedlings by successive extractions followed by anion exchange chromatography−mass spectrometry. Anal. Biochem. 389, 12−17.

Gibon, Y., Vigeolas, H., Tiessen, A., Geigenberger, P., Stitt, M., 2002. Sensitive and high throughput metabolite assays for inorganic pyrophosphate, ADPGlc, nucleotide phosphates, and glycolytic intermediates based on a novel enzymic cycling system. Plant J. 30, 221−235.

Hull, S.R., Montgomery, R., 1994. Separation and analysis of 4′-epimeric UDP-sugars, nucleotides, and sugar phosphates by anion-exchange high-performance liquid chromatography with conductimetric detection. Anal. Biochem. 222, 49−54.

Ikeguchi, Y., Nakamura, H., Nakajima, T., 1993. Organic phosphates analyzer using high-performance anion-exchange chromatography and a postcolumn. Anal. Sci. 9, 653−655.

Jeong, J.S., Kwon, H.J., Lee, Y.M., Yoon, H.R., Hong, S.P., 2007. Determination of sugar phosphates by high-performance anion-exchange chromatography coupled with pulsed amperometric detection. J. Chromatogr. A. 1164, 167−173.

Lefebvre, M.J., Gonzalez, N., Pontis, H.G., 1964. Anion-exchange chromatography of sugar phosphates with triethylammonium borate. J. Chromatogr. 15, 495−500.

Leloir, L.F., 1951. Sugar phosphates. In: Zechmeister, L. (Ed.), Progress in the Chemistry of Organic Natural Products. Springer, Vienna, pp. 47−95.

Leloir, L.F., Cardini, C.E., 1963. Sugar phosphates. In: Florkin, M., Stotz, E.H. (Eds.), Comprehensive Biochemistry, vol. 5. Elsevier Publishing Company, Amsterdam, pp. 113−145.

Mopper, K., Dawson, R., Liebezeit, G., Hansen, H.P., 1980. Borate complex ion exchange chromatography with fluorimetric detection for determination of saccharides. Anal. Chem. 52, 2018−2022.

Murata, R., 1976. Purification and some properties of phosphamannomutase from corms of *Amorphophallus konjac* C. Koch. Plant Cell Physiol. 17, 1099−1109.

Penn, S.G., Hu, H., Brown, P.H., Lebrilla, C.B., 1997. Direct analysis of sugar alcohol borate complexes in plant extracts by matrix-assisted laser desorption/ionization Fourier transform mass spectrometry. Anal. Chem. 69, 2471−2477.

Pontis, H.G., Fischer, C.L., 1963. Synthesis of D-fructopyranose 2-phosphate and D-fructofuranose 2-phosphate. Biochem. J. 89, 452−455.

Pontis, H.G., Leloir, L.F., 1972. Sugar phosphates and sugar nucleotides. In: Halmann, M. (Ed.), Analytical Chemistry of Phosphorus Compounds. John Wiley and Sons, Inc., New York, pp. 617−658.

Ramsay, O.B., Pizanis, M.J., 1965. Hydrolysis of α-D-glucosamine-1-phosphate. Arch. Biochem. Biophys. 110, 32−38.

Sabularse, D.C., Anderson, R.L., 1981. Inorganic pyrophosphate: D-fructose-6-phosphate 1-phosphotransferase in mung beans and its activation by D-fructose 1,6-bisphosphate and D-glucose 1,6-bisphosphate. Biochem. Biophys. Res. Com. 100, 1423−1429.

Sekiguchi, Y., Mitsuhashi, N., Inoue, Y., Yagisawa, H., Mimura, T., 2004. Analysis of sugar phosphates in plants by ion chromatography on a titanium dioxide column with pulsed amperometric detection. J. Chromatogr. A. 1039, 71−76.

Small, D.M., Matheson, N.K., 1979. Phosphomannomutase and phosphoglucomutase in developing *Cassia corymbosa* seeds. Phytochemistry. 18, 1147−1150.

Smrcka, A.V., Jensen, R.G., 1988. HPLC separation and indirect ultraviolet detection of phosphorylated sugars. Plant Physiol. 86, 615−618.

Stitt, M., Lilley, R.M., Gerhardt, R., Heldt, H.W., 1989. Metabolite levels in specific cells and subcellular compartments of plant leaves. Methods Enzymol. 174, 518−552.

Van Schaftingen, E., Lederer, B., Bartrons, R., Hers, H., 1982. A kinetic study of pyrophosphate: fructose-6-phosphate phosphotransferase from potato tubers. Application to a microassay of fructose 2,6-bisphosphate. Eur. J. Biochem. 129, 191−195.

Yamamoto, A., Inoue, Y., Kodama, S., Matsunaga, A., 1999. Capacity gradient anion chromatography with a borate complex as eluent. J. Chromatogr. A. 850, 73−77.

CHAPTER 15

Case Study: Nucleotide Sugars

Chapter Outline

15.1 Introduction

Nucleotide sugars (nucleoside diphosphate sugars, NDP-sugars) are activated sugar donors, energy rich forms of monosaccharides that are ubiquitous in all living organisms. In photosynthetic organisms, NDP-sugars are the donors of the sugar moieties in sucrose, trehalose, oligo-, and polysaccharide synthesis, and are also required for the glycosylation of proteins and lipids, and for the synthesis of a large amount of simple and complex glycosides.

The discovery of uridine diphosphate glucose (UDP-glucose, the most abundant nucleotide sugar in nature) by Leloir and his colleagues (Caputo et al., 1950; Cardini et al., 1950), opened new avenues for the origin and interconversion mechanisms of many of the sugars in living cells. Since that time, the number of sugar nucleotides isolated from animal

Methods for Analysis of Carbohydrate Metabolism in Photosynthetic Organisms.
DOI: http://dx.doi.org/10.1016/B978-0-12-803396-8.00015-6

tissues, plants, and microorganisms has been notably increasing. In plants, there are more than 30 different representatives of these compounds (Feingold and Avigad, 1980; Feingold and Barber, 1990), which include mostly the nucleotides linked to hexoses (D-glucose, D-galactose, D-mannose, and L-galactose) and, to a minor extent, to other sugars such as 6-deoxy hexoses, pentoses, and hexuronic acids (Bar-Peled and O'Neill, 2011). In unicellular algae and cyanobacteria, even though no exhaustive study on NDP-sugars has been reported, glucose nucleotides are shown as important metabolites (Sanwal and Preiss, 1969; Nakamura and Imamura, 1985; Feingold and Barber, 1990; Iglesias et al., 1991; Kawano et al., 2014).

15.2 General Properties

Basically, NDP-sugars have the same structure: a sugar or a sugar derivative bound through a glycosidic hemyacetalic hydroxyl to the terminal phosphate residue of a nucleoside 5′-diphosphate. They only differentiate on the base and the sugar residue. The general structure of NDP-sugars is shown in Fig. 15.1.

The glycosidic linkage "a" is more easily acid hydrolyzed than the "b" and "c" bonds; thus, mild acid hydrolysis conditions yield a nucleoside diphosphate and a free sugar. Also, compounds that have a pyrophosphate group bound though a glycosidic linkage to a sugar are very labile to acid. For example, NDP-sugars such as uridine diphosphate glucose and guanosine diphosphate mannose are more acid-labile than the corresponding sugar 1-phosphate. In general terms, the sugar moiety of the NDP-sugar is released by heating at 100°C for 10–15 min in an acid environment (pH 2). The rate of hydrolysis is influenced by the group type at carbon 6 and the more bulky the group, the more stable the compound is. For example, uridine diphosphate glucose is less stable than uridine diphosphoglucuronic acid.

The pyrophosphate ether bond "b" present in a NDP-sugar can be split under three different conditions: (1) it can be specifically broken by the action of a pyrophsophatase, yielding the corresponding nucleoside monophosphate and a sugar phosphate; (2) the nucleoside

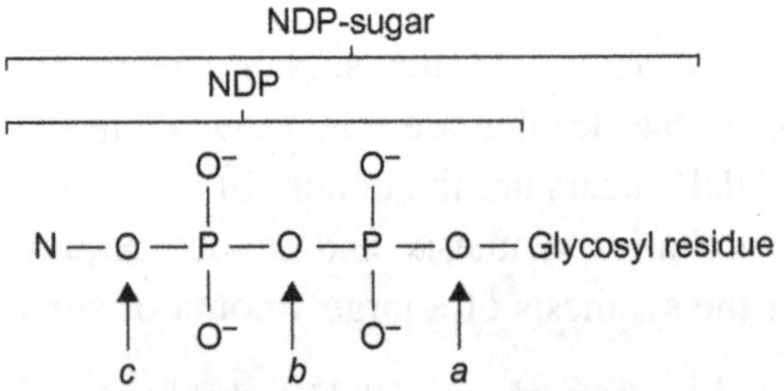

NDP-sugar

NDP

N—O—P(O⁻)(O⁻)—O—P(O⁻)(O⁻)—O—Glycosyl residue

c b a

Figure 15.1

Nucleoside-diphosphate sugar (NDP-sugar) structure. N stands for different nucleosides (U, uridine; A, adenosine; G, guanosine; T, thymidine; C, cytidine). "a," glycosidic bond, which is the most easily split by acid; "b," pyrophosphate bond; and "c," ester bond.

monophosphate can also be obtained under controlled strong acid conditions of hydrolysis (1 N acid at 100°C), which also releases the sugar; and (3) it is also possible to split the pyrophosphate bond under alkaline hydrolysis conditions; however, the products obtained vary according to the sugar moiety nature. For example, when there is no hydroxyl group attached to carbon 2 (such as in acetylhexosamines), the alkaline hydrolysis of the NDP-sugar liberates a nucleoside monophosphate and a sugar phosphate. On the contrary, when the sugar moiety has an hydroxyl group at position *cis* in carbon 2 (like in glucose), the NDP-sugar is very sensitive to alkaline hydrolysis, producing a nucleoside monophosphate and a cyclic sugar 1,2-monophosphate.

The ester bond "c" that links the nucleoside with the phosphate group, is the most resistant to acids, since the purine nucleotide is easier to hydrolyze than the pyrimidine ones.

From a thermodynamic point of view, NDP-sugars are activated sugars, even better donors than sucrose and glucose-1-phosphate. They have a higher free energy of hydrolysis for the glucose group ($\Delta G°' > -7$ Kcal.mol^{-1} at pH $\sim$7.4) than that of α-D-glucose-1-phosphate ($\Delta G°' = -4.8$ Kcal.mol^{-1} at pH 8.5) or sucrose ($\Delta G°' = -6.6$ Kcal.mol^{-1} at pH 8.5) and, consequently, they are better precursors for glycosidic linkages (Leloir et al., 1960; Pontis and Leloir, 1972; Bar-Peled and O'Neill, 2011).

15.3 Biosynthesis of Nucleotide Sugars

In the plant cell, the carbon source for the formation of the diverse NDP-sugars may originate from different pathways: (1) the photosynthesis process yields fructose-6-phosphate as a Calvin-cycle product, which contributes to direct NDP-sugars formation and through interconversion reactions; (2) sucrose mobilization by the action of sucrose synthase cleaves sucrose to NDP-glucose and fructose; (3) the mobilization of storage polysaccharide such as starch or glucomannans; (4) the recycling of sugar residues (ie, from glycans such as glycolipids or glycoproteins, secondary metabolites or polysaccharides constituents of the primary and secondary cell walls); and (5) plant–microbe interactions (Bar-Peled and O'Neil, 2011). D-fructose-6-phosphate withdrawn from the Calvin cycle is the precursor of D-mannose-1-phosphate, D-glucose-1-phosphate, and acetylglucamine-1-phosphate, which are in turn converted to the corresponding NDP-sugar. Thus, fructose-6-phosphate is the main precursor for the formation of the major plant NDP-sugars. On the other hand, other monosaccharides resulting from recycling or salvage pathways are phosphorylated by specific kinases, or enter a nucleotide pool without modification (Feingold and Avigad, 1980).

In general, the biosynthesis of NDP-sugars involves the transfer of a nucleotidyl group from a nucleoside triphosphate (NTP) to a sugar-1-phosphate, with the simultaneous release of pyrophosphate (PPi), reaction catalyzed by pyrophosphorylase enzymes:

$$\text{NTP} + \text{sugar-1-phosphate} \xleftrightarrow{\text{Pyrophosphorylase}} \text{PPi} + \text{NDP-sugar}$$

where N could be uridine (U), adenosine (A), cytidine (C), guanosine (G), or thymidine (T). This reaction is easily reversible in the presence of Mg^{2+}. In vivo, the reaction is shifted to the right due to the action of inorganic pyrophosphatases that hydrolyze the pyrophosphate. Pyrophosphorylases are specific for the nucleoside triphosphate. The sugar moiety can be glucose, mannose, fucose, acetylglucosamine, xylose, arabinose, or glucuronic acid. UDP-glucose and GDP-D-mannose can be produced directly from D-fructose-6-phosphate. Other NDP-sugars are formed after different epimerization, decarboxylation or dehydrogenation reactions. GDP-L-galactose and GDP-L-fucose are produced from GDP-D-mannose, and UDP-galactose, UDP-D-glucuronic acid, and UDP-D-xylose derive from UDP-D-glucose (Yin et al., 2011). Both UDP-galactose and UDP-galacturonic are formed by an epimerization of the hydroxyl group in the position 4 of the sugar moiety of UDP-glucose and UDP-glucuronic acid, respectively. Additionally, NDP-sugars can be directly produced from free sugars through alternative pathways.

The knowledge of nucleotide sugar interconversion pathways in unicellular algae and cyanobacteria is very scarce and generally limited to the computational identification of sequences homologous to coding genes of characterized enzymes in other organisms (eg, in *Chlamydomonas reinhardtii*) (Yin et al., 2011).

Glucose-containing oligo- and polysaccharides are in the majority in oxygenic photosynthetic organisms, and their syntheses involve the availability of ADP-glucose, GDP-glucose, or UDP-glucose. The activity of pyrophosphorylases that catalyze the formation of glucose-containing NDP-sugars can be assayed by various methods. Usually, the activity is measured as the NDP-sugar pyrophosphorolysis reaction (reverse direction) using NDP-sugar and PPi, and determining glucose-1-phosphate formation. In the presence of phosphoglucomutase and glucose-6-phosphate dehydrogenase the reaction can be followed by NADPH formation. Another method uses radioactive pyrophosphate, with the concomitant formation of radioactive nucleoside triphosphate. The latter is separated from the remaining pyrophosphate by adsorption in activated carbon.

The NDP-sugar formation reaction (forward) can be assayed using an α-phosphate-labeled nucleoside triphosphate. This method can be applied for the determination of any NDP-sugar formation. The fourth general procedure uses labeled glucose-1-phosphate as substrate. The determination of the labeled nucleotide sugar produced in the reaction is achieved after incubation with a pyrophosphatase-free alkaline phosphatase that hydrolyzes the unreacted glucose-1-phosphate. Radioactive glucose is separated from the nucleotide sugar formed by passing the solution through a Dowex-1-Cl^- column (equilibrated at pH 8), where only the nucleotide sugar is retained. The latter is eluted from the anionic resin with a 1 M NaCl solution at pH 2 and radioactivity in the eluate is measured in a scintillation counter.

Experimental Procedures

15.4 Extraction of Nucleotide Sugars

Nucleotide sugars represent 10–25% of the total pool of soluble nucleotides extracted and in most plant material studies the uridine nucleotides (mainly UDP-glucose) predominate (Feingold and Avigad, 1980). The extraction conditions from plant tissues or cells must involve minimal nucleotide degradation and avoid contamination by interfering substances (Feingold and Barber, 1990). Early methods involved extraction with 80% (v/v) hot ethanol (Sanwal and Preiss, 1969). However, it was shown that the hot solution activates various phosphatases that degrade the labile phosphorylated NDP-sugars. To obtain stable NDP-sugars preparations, further studies used cold ethanol acidified with formic acid, which can be removed by vacuum sublimation. A classical protocol starts with frozen tissue in liquid nitrogen that is homogenized in a blender with cold 80% formic acid and poured into ethanol (50 volumes) at −20°C for 24 h (to inactivate phosphatases). The extract is brought to pH 6.5 with ammonium hydroxide, and the supernatant of centrifugation is reduced in volume in vacuum. The aqueous solution obtained is freed from ammonium formate by lyophilization. The residue is dissolved in water and ready for further purification with classical methods (Feingold and Barber, 1990). Nucleotides and NDP-sugars mixtures extracted with cold formic acid must be pretreated with charcoal before separation procedures in order to eliminate other compounds containing terminal phosphate groups. The nucleotides eluted from the charcoal are then incubated with phosphomonoesterase to eliminate nucleoside mono-, di-, and triphosphates (which are converted to nucleosides and inorganic phosphate). Then, NDP-sugars can be separated from those low molecular mass compounds by gel chromatography (Feingold and Barber, 1990).

The use of high-performance liquid chromatography (HPLC) demands other special extraction techniques, such as that developed by Meyer and Wagner (1985). For metabolomic analyses, metabolite extraction is usually carried out with trichloroacetic acid (Jelitto et al., 1992). Two frequently used extraction methods are described below.

15.4.1 Extraction With Perchloric Acid

Principle

This procedure is used for the extraction of nucleotides and NDP-sugars from plant tissues for further determination by HPLC. It involves four steps: extraction with perchloric acid, purification on a disposable prepacked phenyl-bonded column, neutralization, and concentration. The method was applied to *Nicotiana tabacum* cells and to other plant tissues such as leaves and roots (Meyer and Wagner, 1985).

Reagents

Biological material: nitrogen-liquid frozen tissue or cells	
$HClO_4$ (perchloric acid)	0.4 M
Phenylsilane bonded silica gel	
Methanol	
KOH solutions for neutralization	5 M and 0.1 M
HCl	10 mM

Note: Perform all procedures at 4°C or in ice.

Procedure

Grind the frozen tissue (or cells) in a mortar cooled with liquid nitrogen. After weighing, transfer the material into a Potter–Elvejhem glass homogenizer. Add chilled 0.4 M $HClO_4$ (2 mL–0.4 g of frozen material). After homogenization, incubate at 0–4°C for 10 min and centrifuge at 39,000 × g for 10 min. Prepare a column (eg, a disposable pipette) of phenylsilane bonded silica gel (for 0.4 g of tissue, use a 3 mL disposable pipette containing 500 mg of silica gel). Wash the column with 1 volume of methanol, at a rate of c.3 $mL \cdot min^{-1}$ using a peristaltic pump. Equilibrate the column with 2 mL of 0.4 M $HClO_4$. Apply to the column 0.5 mL of the supernatant obtained after centrifugation and elute with 0.4 M $HClO_4$. Collect 4 mL of eluate, which will be carefully neutralized with cold KOH to a pH between 5.5 to 6.5 (use more concentrated KOH solution until reaching pH ~5.5, and then use a less concentrated alkali solution). Remove the potassium perchlorate precipitate by centrifugation and lyophilize the supernatant to dryness. Dissolve the dry material in 250 μL of 10 mM HCl, held at 0–4°C for some minutes and then centrifuge to remove traces of potassium perchlorate. Store the supernatant solution at −70°C until use.

15.4.2 Extraction With Trichloroacetic Acid

Principle

The extraction of metabolites, including NDP-sugars, with trichloroacetic acid (Weiner et al., 1987) is a convenient technique to obtain samples for instrumental analyses (Jelitto et al., 1992).

Reagents

Trichloroacetic acid (TCA)	16% in diethylether (v/v)
TCA solution	16% in water (w/v) containing 5 mM NaF
Water-saturated ether	
5 M KOH/1 M triethanolamine mixture	
Active charcoal	

Note: Prewash mortar and all materials for 12 h in 2 N HCl.

Procedure

Homogenize 200–300 mg of frozen plant material to a fine powder in a mortar (previously cooled with dry ice) under liquid nitrogen. Add 1.5 mL of 16% TCA in diethylether (v/v), precooled to the temperature of dry ice, and homogenize again. After standing for 20 min on dry ice, transfer the mortar onto ice at 4°C and add 0.8 mL of 16% TCA (water solution, w/v) containing 5 mM NaF (to inhibit pyrophosphatases). Hold at 4°C for 3 h to achieve complete denaturation of pyrophosphatases. Centrifuge the extract for 5 min and remove the lower (water) phase. Wash the water phase four times by shaking with 0.8 mL of water-saturated ether and centrifuge again. Neutralize with a small volume of a 5 M KOH/1 M triethanolamine mixture, treated with active charcoal. Freeze-dry in liquid nitrogen until use.

15.5 Determination of Sugar Nucleotides

Chemical analysis of NDP-sugars may be carried out by determining the sugar, or the nucleoside moiety, or the phosphate groups. The sugar can be easily identified and quantified after mild acid hydrolysis (pH 2) at 100°C for 5–15 min using one of the methods described in Chapter 1, Determination of Carbohydrates Metabolism Molecules. Nucleosides have typical absorption spectra in neutral, acid, and alkaline solutions and their concentrations can be determined spectrophotometrically using the molar absorbance coefficient for each nucleoside. At pH 7, λ_{max} for adenosine, cytidine, guanosine, and thymidine are 259, 280, 252, and 262 nm, respectively, and the molar absorbance coefficient are 15.4×10^3, 13.0×10^3, 13.7×10^3, and 10.2×10^3, respectively.

To ascertain the presence of NDP after release of the sugar component, different techniques can be used (paper chromatography, electrophoresis, or HPLC). Also, after acid treatment (1 N acid at 100°C for 15 min), the liberation of one phosphate group (the second phosphate group is resistant to this acid hydrolysis) can be estimated as described in Chapter 1, Determination of Carbohydrates Metabolism Molecules.

On the other hand, specific enzymatic reactions can be used for determining NDP-sugars. In Chapter 1, Determination of Carbohydrates Metabolism Molecules, the determination of UDP-glucose using UDP-glucose dehydrogenase or UDP-glucose pryrophosphorylase are described (Pontis and Leloir, 1962). A similar protocol can be employed for ADP-glucose determination using ADP-glucose pyrophosphorylase (Espada, 1966).

15.6 UDP-Glucose Pyrophosphorylase Activity Assays

UDP-glucose pyrophosphorylase (EC 2.7.7.9) represents an important activity in photosynthetic carbohydrate metabolism, catalyzing the reversible reaction:

$$\text{UTP} + \text{glucose-1-phosphate} \leftrightarrow \text{UDP-glucose} + \text{PPi}$$

In plants UDP-glucose pyrophosphorylase is present at high activity and occurs as at least two isozymes. In source tissues, the enzyme provides UDP-glucose for sucrose and cell wall biosynthesis (Kleczkowski et al., 2004).

15.6.1 Enzymatic Assay (Reverse Direction)

Principle

Enzyme activity can be assayed in the UDP-glucose pyrophosphorolysis direction. Glucose-1-phosphate formed is determined by coupling phosphoglucomutase and glucose-6-phosphate dehydrogenase, and measuring NADPH formation at 340 nm (Pontis and Leloir, 1962).

$$\text{Glucose-1-phosphate} \xleftrightarrow{\text{Phosphoglucomutase + coenzyme}} \text{Glucose-6-phosphate}$$

$$\text{Glucose-6-phosphate} + \text{NADP}^+ \xrightarrow{\text{Glucose-6-phosphate dehydrogenase}} \text{6-Phosphogluconate} + \text{NADPH} + \text{H}^+$$

Reagents

Tris-HCl buffer (pH 7.2)	50 mM containing $MgCl_2$ 5 mM
Cysteine (neutralized)	$10\ \text{mg} \cdot \text{mL}^{-1}$
UDP-glucose	7 mM
Sodium pyrophosphate	0.1 M
$NADP^+$	0.05 M
Glucose-1,6-diphosphate	2 mM
Phosphoglucomutase (from yeast)	$3\ \text{mg} \cdot \text{mL}^{-1}$
Glucose-6-phosphate dehydrogenase (from yeast)	

Procedure

Prepare the reaction mixture in a 1-mL cuvette with 1-cm light path. In a 1-mL total volume, mix 800 μL of Tris-HCl buffer, 25 μL of phosphoglucomutase ($3\ \text{mg} \cdot \text{mL}^{-1}$), 25 μL of cysteine, 10 μL of glucose-6-phosphate dehydrogenase (UDP-glucose pyrophosphorylase free), 50 μL of UDP-glucose, 10 μL of $NADP^+$, 10 μL glucose-1,6-diphosphate, and an aliquot of the enzyme preparation. Place the cuvette in a suitable thermostatted spectrophotometer (at 25°C). Start the reaction with the addition of 10 μL of PPi. Register the increase in absorbance at 340 nm every minute until constant is observed.

To calculate the μmoles of hexose phosphate present in the sample, divide the increase in absorbance per unit of time by 6.22, taking into account that the absorbance at 340 nm of 1 mL (in a 1-cm path length cuvette) containing 1 μmol of NADPH is 6.22.

Comment

The method described above is applicable to purified or pure extracts of UDP-glucose pyrophosphorylase. For crude extracts, the reaction is carried out in test tubes and in two steps. First, the reaction mixture is prepared in a final volume of 100 μL, omitting PPi, $NADP^+$, the two auxiliary enzymes, and the cofactor glucose-1,6-diphosphate. After addition of an aliquot of the enzyme preparation, the reaction is started with PPi and incubated at least three times, eg, at 2, 4, and 10 min. The reaction is stopped by heating at 100°C for 2 min. Once cooled, the auxiliary enzymes, $NADP^+$, and glucose-1,6-diphosphate are added to the tubes and the mixture is made up to 1 mL. After incubation at 30°C for 10 min, NADPH formation is determined at 340 nm.

15.6.2 Enzymatic Assay (Forward Direction)

Principle

Enzyme activity can be assayed in the UDP-glucose synthesis direction by colorimetrically quantifying the formation of orthophosphate (Pi) after PPi hydrolysis by the addition of an inorganic pyrophosphatase.

Reagents

MOPS (morpholinepropanesulfonic acid) (pH 8.0)	500 mM
$MgCl_2$	100 mM
UTP	20 mM
Bovine serum albumin (BSA)	$0.2\ mg \cdot mL^{-1}$
Inorganic pyrophosphatase (from yeast)	
Glucose-1-phosphate	10 mM

Procedure

In a 50-μL total volume, mix 5 μL of MOPS buffer (pH 8.0), 5 μL of $MgCl_2$, 5 μL of UTP, 0.01 mg of BSA, 0.0005 unit $\cdot\ \mu L^{-1}$ of yeast inorganic pyrophosphatase, and the appropriate enzyme dilution. Start the reaction by the addition of 5 μL of glucose-1-phosphate. Incubate at 37°C for 10 min. Stop the reaction by the addition of the reagent for inorganic phosphate determination (see Chapter 1: Determination of Carbohydrates Metabolism Molecules).

Comments

The method described in this section is used to assay activity of purified enzymes. Also it is applicable to assay ADP-glucose pyrophosphorylase, using ATP (instead UTP) as substrate.

The reaction can be alternatively terminated with the malachite green reactive (Fusari et al., 2006) and the complex formed with the released Pi is measured at 630 nm with an ELISA (enzyme-linked immunosorbent assay) detector (Asención Diez et al., 2012). Another enzyme assay using D-[1-^{13}C]glucose-1-phsophate and UTP was used to determine UDP-glucose phyrophosphorylase activity in chloroplasts (Okazaki et al., 2009). The reaction is stopped with methanol and the amount of UDP-glucose formed is quantified by UPLC-MS analysis.

15.7 ADP-Glucose Pyrophosphorylase Activity Assays

ADP-glucose pyrophosphorylase (EC 2.7.7.27) occupies a central role in the plant life, catalyzing the reversible reaction:

$$\text{ATP} + \text{glucose-1-phosphate} \leftrightarrow \text{ADP-glucose} + \text{PPi}$$

In plants ADP-glucose pyrophosphorylase is a highly regulated enzyme directly involved in starch synthesis because it generates the glucosyl donor (ADP-glucose) for the elongation of α-1,4-glucosidic chains (Ballicora et al., 2004).

15.7.1 Enzymatic Assay (Pyrophosphorolysis Direction)

Principle

An alternative of the method described in Section 15.6.1 for UDP-glucose pyrophosphorylase, based on the same principle, can be used for the assay of ADP-glucose pyrophosporylase. The glucose-1-phosphate formed is determined by coupling phosphoglucomutase and glucose-6-phosphate dehydrogenase, and measuring NADPH formation at 340 nm.

$$\text{Glucose-1-phosphate} \xleftrightarrow{\text{Phosphoglucomutase}} \text{Glucose-6-phosphate}$$

$$\text{Glucose-6-phosphate} + \text{NADP}^+ \xrightarrow{\text{Glucose-6-phosphate dehydrogenase}} \text{6-Phosphogluconate} + 2\ \text{NADPH}$$

The amount of the reduced product (NADPH) is proportional to the quantity of glucose-1-phosphate (Pontis and Leloir, 1962).

Reagents

Tris-HCl buffer (pH 7.8)	0.5 M
$MgCl_2$ (magnesium chloride)	0.05 M
ADP-glucose	10 mM
Sodium pyrophosphate (PPi)	0.02 M
$NADP^+$	0.05 M
Phosphoglucomutase	$100\ U \cdot mg\ protein^{-1}$
Glucose-6-phosphate dehydrogenase (from yeast)	$200\ U \cdot mg\ protein^{-1}$

Procedure

In a total volume of 100 μL, mix 10 μL of Tris-HCl buffer (pH 7.8), 10 μL of $MgCl_2$, 10 μL of ADP-glucose, and an aliquot of the enzyme preparation. Start the reaction by adding 10 μL of PPi and incubate at 30°C for 10 min. Stop the reaction by heating at 100°C for 2 min. Once cooled, add the auxiliary enzymes (phosphoglucomutase and glucose-6-phosphate dehydrogenase, appropriately diluted) and 10 μL of $NADP^+$. Make up to 1 mL and incubate at 25–30°C in a 1-cm path cuvette. Register the absorbance at 340 nm at 1-min intervals until no further reaction is detected. Run a control without pyrophosphate. Calculate the μmoles of glucose-1-phosphate in the mixture, dividing the increase in absorbance at 340 nm by 6.22.

15.7.2 Radioactive Assay (Pyrophosphorolysis Direction)

Principle

Enzyme activity is assayed by measuring the ADP-glucose pyrophosphorolysis by the formation of [^{32}P]ATP from ^{32}PPi. The ATP formed is isolated by adsorption over an activated charcoal (Norit A) and its estimation involves the determination of radioactivity contained in the charcoal (Ghosh and Preiss, 1966; Fu et al., 1998; Ballicora et al., 2000).

Reagents

MOPS buffer (pH 8.0)	0.5 M
ADP-glucose	10 mM
^{32}P-Sodium pyrophosphate (^{32}PPi)	0.01 M (specific activity $0.5-2 \times 10^6$ cpm · μmol^{-1})
$MgCl_2$	0.1 M
NaF	0.20 M
Bovine serum albumin (BSA)	
Trichloroacetic acid (TCA)	10% (v/v) and 5% (v/v)
Norit A suspension (charcoal)	150 mg · mL^{-1}
Ethanol	50% solution containing 0.1% NH_3

Procedure

In a total volume of 150 μL, mix 15 μL of ADP-glucose, 15 μL of MOPS buffer (pH 8.0), 8 μL $MgCl_2$ and 8 μL NaF, 0.2 mg · mL^{-1} of BSA, and an aliquot of the enzyme preparation. Start the reaction with the addition of 15 μL of ^{32}PPi. Incubate at 37°C for 10 min. Stop the reaction by adding 1 mL of cold 10% TCA. The [^{32}P]ATP formed was measured following Shen and Preiss (1964). In order to dilute radioactive PPi, add 100 μmoles of nonlabeled PPi and add 100 μL Norit A suspension to absorb radioactive ATP. Centrifuge the Norit suspension and discard the supernatant. Wash the charcoal (Norit A precipitate) with 3 mL cold TCA 5% and then once with 3 mL cold distilled water. After the washings, resuspend Norit A in 2 mL of 50% ethanol solution containing 0.1% NH_3. One milliliter of this suspension is dried on an aluminum planchet and counted in a gas-flow counter.

Comments

This method can also be used to assay UDP-glucose pyrophosphorylase and GDP-mannose pyrophosphorylase activity, replacing ADP-glucose by UDP-glucose (Asención Diez et al., 2012) or GDP-mannose (Preiss, 1966), respectively.

15.8 GDP-Mannose Pyrophosphorylase Activity Assay

In plants, the reaction catalyzed by the enzyme GDP-mannose pyrophosphorylase (EC 2.7.7.13):

$$\text{GTP} + \text{Mannose-1-phosphate} \leftrightarrow \text{GDP-mannose} + \text{PPi}$$

is part of the major L-galactose pathway that leads to L-ascorbate biosynthesis, and is also related with glycoproteins and cell wall polysaccharides (Linster and Clarke, 2008; Barth et al., 2010).

Principle

The enzyme is assayed by measuring the rate of formation of GDP-[^{14}C]-mannose from [^{14}C]-mannose-1-phosphate and GTP. The procedure involves the hydrolysis of the unreacted [^{14}C]-mannose-1-phosphate with the addition of an alkaline phosphatase (Preiss, 1966).

Reagents

[^{14}C]-mannose-1-phosphate	10 mM (specific activity $2–20 \times 10^5$ cpm · μmol^{-1})
GDP	10 mM
Tris-HCl buffer (pH 8)	50 mM, containing 10 mM of $MgCl_2$ and 1 mM of EDTA
NaCl solution	1 M NaCl in 0.01 N HCl
Crystalline alkaline phosphatase (pyrophosphatase-free)	
Dowex-1-Cl^- or AG1-X4 (chloride form) 200-400 mesh	

Procedure

In a 200-μL total volume, mix 80 μL of the Tris-HCl buffer solution (pH 8.0), 10 μL of [^{14}C]-mannose-1-phosphate, 10 μL of GTP, and an aliquot of the enzyme preparation.

Incubate at 30°C for 15 min. Stop the reaction by heating the mixture at 100°C for 1 min. After cooling, add 5 μL of alkaline phosphatase (10 μg, 10 U · mg^{-1} of protein) to the reaction mixture and incubate at 37°C for 30 min in order to hydrolyze the unreacted labeled mannose-1-phosphate. Stop the reaction by adding 2 mL of cold water (ice temperature). Filter the mixture through a AG1-X4 (chloride form) column (0.6 × 2 cm). Wash the column twice with 2 mL of distilled water. Elute the GDP-[^{14}C]-mannose formed stick to the column with 2 mL of NaCl solution (1 M in 0.01 N HCl). Collect the eluate in a vial, add scintillation liquid, and determine the radioactivity in a scintillation counter.

15.9 UDP-Sugar-4-Epimerase Activity Assay

UDP-glucose 4-epimerase (EC 5.1.3.2) freely interconverts UDP-glucose and UDP-galactose.

$$\text{UDP-galactose} \leftrightarrow \text{UDP-glucose}$$

Different isoforms occur in plants that play a major role in vegetative growth and in the regulation of cell wall carbohydrate biosynthesis, and also cooperate in pollen development (Dörmann and Benning, 1998; Rösti et al., 2007).

Principle

The epimerase activity is followed spectrophotometrically at 340 nm by measuring UDP-glucose formation with UDP-glucose dehydrogenase and NADH (Pontis and Leloir, 1962).

Reagents

UDP-galactose	7 mM
Glycine buffer (pH 9.0)	1 M
UDP-glucose dehydrogenase	
NAD^+	0.05 M

Procedure

In a 0.5-mL total volume, mix 5 μL of UDP-galactose, 10 μL of NAD^+, 50 μL of glycine buffer (pH 9.0), and 200 units of UDP-glucose dehydrogenase. Incubate for several minutes at 25°C after addition of the dehydrogenase (UDP-galactose preparations usually contains UDP-glucose). When a constant absorbance at 340 nm is reached, add an aliquot of the UDP-galactose-4-epimerase preparation and register the absorbance for 4 min. Run a blank without UDP-galactose.

A unit of activity is defined as an increase of 0.001 min^{-1} in absorbance at 340 nm under the assay conditions. A change in absorbance at 340 nm of 12.4 corresponds to 1 $\mu mol \cdot mL^{-1}$ of UDP-glucose formed.

15.10 UDP-Glucoronic Acid Decarboxylase Activity Assay

UDP-D-glucuronic decarboxylase (EC 4.1.1.35) catalyzes the synthesis of UDP-D-xylose from UDP-D-glucuronic acid in an essentially irreversible reaction that is believed to commit glycosyl residues to heteroxylan and xyloglucan biosynthesis.

$$\text{UDP-D-glucoronic acid} \rightarrow \text{UDP-D-xylose} + CO_2$$

Thus, it is a key enzyme in the synthesis of the precursor (UDP-xylose) for the formation of xylans during cell wall biosynthesis (Zhang et al., 2005; Du et al., 2013).

Principle

UDP-D-xylose produced from UDP-glucuronic acid is hydrolyzed in an acid medium, releasing xylose, which is determined colorimetrically (Ankel and Feingold, 1966).

Reagents

UDP-D-glucoronic acid	0.03 M
Na and K phosphate buffer (pH 7.0)	0.1 M, containing 0.5 g EDTA and 0.5 mL of β-mercapthoethanol per liter
Glacial acetic acid	

Note: All the reagents from the enzymatic reaction must be at pH 7. Reagents should be at 37°C before preparing the reaction mixture.

Procedure

In a 300-μL total volume, mix 40 μL of UDP-D-glucuronic acid, 260 μL of buffer, and an aliquot of the enzyme preparation. Incubate at 37°C. At appropriate times, stop the reaction by transferring 0.1-mL samples to small tubes containing 60 μL glacial acetic acid. Cover the tubes with marbles and heat at 100°C for 15 min in order to hydrolyze the sugar nucleotides. Determine the D-xylose formed by the orcinol method described in Chapter 1, Determination of Carbohydrates Metabolism Molecules. A standard curve must be run with each assay, using quantities of D-xylose ranging from 0.05 to 0.03 $\mu mol.sample^{-1}$.

15.11 Other Enzyme Activity Assays

Thymidine diphosphate-glucose and cytidine diphosphate-glucose are usually determined as has been previously described for UDP-glucose and ADP-glucose, using the auxiliary enzymes phosphoglucomutase, glucose-6-phosphate dehydrogenase, and NADP.

15.12 Overview of Nucleotide-Sugars Separation Methods

NDP-sugars can be separated and identified either by classical or instrumental methods. Paper chromatography and paper electrophoresis have been extensively employed for separation and identification (Pontis and Leloir, 1972). Also, thin layer chromatography for analytical purposes is a technique that offers several advantages over paper chromatography, eg, higher sensitivity (0.1 μg of unlabeled nucleotide can be detected) and faster development (Feingold and Barber, 1990).

The most convenient method for fractionation of NDP-sugars by low pressure liquid chromatography is using anion-exchange resins (eg, Dowex-1 in the chloride or formate form). The sample (a diluted neutral solution) is adsorbed and then eluted with a salt gradient (eg, with sodium chloride or sodium formate for a chloride or formate form resin, respectively). NDP-sugars are detected by absorbance at 260 nm. The order in which the compounds elute from the column depends both on the anions and on the cations of the displacing solution. The eluted compounds are concentrated by charcoal adsorption–desorption (Pontis and Leloir, 1972; Feingold and Barber, 1990).

On the other hand, NDP-sugars separation can be successfully achieved by HPLC. The use of anion-exchange and reversed-phase HPLC allows the examination and determination of a wide spectrum of NDP-sugars (Meyer and Wagner, 1985). HPLC methodology is in constant improvement and novel uses have been found to analyze NDP-sugars (Hull and Montgomery, 1994; Goulard et al., 2001; Yang and Bar-Peled, 2010). More recently, highly-sensitive chromatographic methods using special equipment have been developed. These can be applied for the resolution and quantitation of NDP-sugars (eg, high-performance anion-exchange chromatography coupled to electrochemical detection or liquid chromatolinked to mass spectrometry) (Arrivault et al., 2009; Nakajima et al., 2010; Pabst et al., 2010; Behmüller et al., 2014; Ito et al., 2014).

Further Reading and References

Ankel, H., Feingold, D.S., 1966. UDP-glucuronic acid decarboxylase. In: Neufeld, E.F., Ginsburg, V. (Eds.), Methods in Enzymology, vol. VIII. Academic Press, New York, pp. 287–290.

Arrivault, S., Guenther, M., Ivakov, A., Feil, R., Vosloh, D., Van Dongen, J.T., et al., 2009. Use of reverse-phase liquid chromatography, linked to tandem mass spectrometry, to profile the Calvin cycle and other metabolic intermediates in Arabidopsis rosettes at different carbon dioxide concentrations. Plant J. 59, 826–839.

Asención Diez, M.D., Peirú, S., Demonte, A.M., Gramajo, H., Iglesias, A.A., 2012. Characterization of recombinant UDP- and ADP-glucose pyrophosphorylases and glycogen synthase to elucidate glucose-1-phosphate partitioning into oligo- and polysaccharides in *Streptomyces coelicolor*. J. Bacteriol. 194 (6), 1485–1493.

Ballicora, M.A., Frueauf, J.B., Fu, Y., Schürmann, P., Preiss, J., 2000. Activation of the potato tuber ADP-glucose pyrophosphorylase by thioredoxin. J. Biol. Chem. 275, 1315–1320.

Ballicora, M.A., Iglesias, A.A., Preiss, J., 2004. ADP-glucose pyrophosphorylase: a regulatory enzyme for plant starch synthesis. Photosynth. Res. 79, 1–24.

Bar-Peled, M., O'Neill, M.A., 2011. Plant nucleotide sugar formation, interconversion, and salvage by sugar recycling. Annu. Rev. Plant. Biol. 62, 127–155.

Barth, C., Gouzd, Z.A., Steele, H.P., Imperio, R.M., 2010. A mutation in GDP-mannose pyrophosphorylase causes conditional hypersensitivity to ammonium, resulting in Arabidopsis root growth inhibition, altered ammonium metabolism, and hormone homeostasis. J. Exp. Bot. 61, 379–394.

Behmüller, R., Forstenlehner, I.C., Tenhaken, R., Huber, C.G., 2014. Quantitative HPLC-MS analysis of nucleotide sugars in plant cells following off-line SPE sample preparation. Anal. Bioanal. Chem. 406, 3229–3237.

Caputo, R., Leloir, L.F., Cardini, E.C., Paladini, A., 1950. Isolation of the coenzyme of the galactose phosphate–glucose phosphate transformation. J. Biol. Chem. 184, 333–350.

Cardini, C.E., Paladini, A.C., Caputo, R., Leloir, L.F., 1950. Uridine diphosphate glucose: the coenzyme of the galactose–glucose phosphate isomerization. Nature. 165, 191–192.

Dörmann, P., Benning, C., 1998. The role of UDP-glucose epimerase in carbohydrate metabolism of *Arabidopsis*. Plant J. 13, 641–652.

Du, Q., Pan, W., Tian, J., Li, B., Zhang, D., 2013. The UDP-glucuronate decarboxylase gene family in populus: structure, expression, and association genetics. PLoS One. 8 (4), e60880. Available from: http://dx.doi.org/10.1371/journal.pone.0060880/. (accessed 10.02.16).

Espada, J., 1966. ADP-glucose pyrophosphorylase from corn grain. In: Neufeld, E.F., Ginsburg, V. (Eds.), Complex Carbohydrates. Methods in Enzymology, vol. VIII. Academic Press, New York and London, pp. 259–262.

Feingold, D.S., Avigad, G., 1980. Sugar nucleotide transformations in plants. In: Preiss, J. (Ed.), The Biochemistry of Plants: A Comprehensive Treatise, vol. 3. Academic Press, New York, pp. 101–170.

Feingold, D.S., Barber, G.A., 1990. Nucleotide sugars. In: Dey, P.M. (Ed.), Carbohydrates, Methods in Plant Biochemistry, vol. 2, pp. 39–78.

Fu, Y., Ballicora, M.A., Leykam, J.F., Preiss, J., 1998. Mechanism of reductive activation of potato tuber ADP-glucose pyrophosphorylase. J. Biol. Chem. 273, 25045–25052.

Fusari, C., Demonte, A.M., Figueroa, C.M., Aleanzi, M., Iglesias, A.A., 2006. A colorimetric method for the assay of ADP-glucose pyrophosphorylase. Anal. Biochem. 352, 145–147.

Ghosh, H.P., Preiss, J., 1966. Adenosine diphosphate glucose pyrophosphorylase. A regulatory enzyme in the biosynthesis of starch in spinach leaf chloroplasts. J. Biol. Chem. 241, 4491–4504.

Goulard, F., Diouris, M., Deslandes, E., Floc'h, J.Y., 2001. An HPLC method for the assay of UDP-glucose pyrophosphorylase and UDP-glucose-4-epimerase in *Solieria chordalis* (Rhodophyceae). Phytochem. Anal. 12, 363–365.

Hull, S.R., Montgomery, R., 1994. Separation and analysis of 4′-epimeric UDP-sugars, nucleotides, and sugar phosphates by anion-exchange high-performance liquid chromatography with conductimetric detection. Anal. Biochem. 222, 49–54.

Iglesias, A.A., Kakefuda, G., Preiss, J., 1991. Regulatory and structural properties of the cyanobacterial ADPglucose pyrophosphorylases. Plant Physiol. 97, 1187–1195.

Ito, J., Herter, T., Baidoo, E.E.K., Lao, J., Vega-Sánchez, M.E., Smith-Moritz, A.M., et al., 2014. Analysis of plant nucleotide sugars by hydrophilic interaction liquid chromatography and tandem mass spectrometry. Anal. Biochem. 448, 14–22.

Jelitto, T., Sonnewald, U., Willmitzer, L., Hajirezaei, M.R., Stitt, M., 1992. Inorganic pyrophosphate content and metabolites in leaves and tubers of potato and tobacco plants expressing *E. coli* pyrophosphatase in their cytosol: biochemical evidence that sucrose metabolism has been manipulated. Planta. 188, 238–244.

Kawano, Y., Sekine, M., Ihara, M., 2014. Identification and characterization of UDP-glucose pyrophosphorylase in cyanobacteria *Anabaena* sp. PCC 7120. J. Biosc. Bioeng. 117, 531–538.

Kleczkowski, L.A., Geisler, M., Ciereszko, I., Johansson, H., 2004. UDP-Glucose pyrophosphorylase. An old protein with new tricks. Plant Physiol. 134, 912–918.

Leloir, L.F., Cardini, C.E., Cabib, E., 1960. In: Florkin, M., Mason, H.S. (Eds.), Comparative Biochemistry, 2. Academic Press, New York, pp. 97–138.

Linster, C.L., Clarke, S.G., 2008. L-ascorbate biosynthesis in higher plants: the role of VTC2. Trends Plant Sci. 13, 567–573.

Meyer, R., Wagner, K.G., 1985. Determination of nucleotide pools in plant tissue by high-performance liquid chromatography. Anal. Biochem. 148, 269–276.

Nakajima, K., Kitazume, S., Angata, T., Fujinawa, R., Ohtsubo, K., Miyoshi, E., et al., 2010. Simultaneous determination of nucleotide sugars with ion-pair reversed-phase HPLC. Glycobiology. 20, 865–871.

Nakamura, Y., Imamura, M., 1985. Regulation of ADP-glucose pyrophosphorylase from *Chlorella vulgaris*. Plant Phys. 78, 601–605.

Okazaki, Y., Shimojima, M., Sawada, Y., Toyooka, K., Narisawa, T., Mochida, K., et al., 2009. A chloroplastic UDP-glucose pyrophosphorylase from Arabidopsis is the committed enzyme for the first step of sulfolipid biosynthesis. Plant Cell. 21, 892–909.

Pabst, M., Grass, J., Fischl, R., Léonard, R., Jin, C., Hinterkörner, G., et al., 2010. Nucleotide and nucleotide sugar analysis by liquid chromatography-electrospray ionization-mass spectrometry on surface-conditioned porous graphitic carbon. Anal. Chem. 82, 9782–9788.

Pontis, H.G., Leloir, L.F., 1962. Measurement of UDP-enzymes systems. In: Glick, D. (Ed.), Methods of Biochemical Analysis, vol. 10. Interscience Publisher. John Wiley and Sons, New York, pp. 107–136.

Pontis, H.G., Leloir, L.F., 1972. Sugar phosphates and sugar nucleotides. In: Halmann, M. (Ed.), Analytical Chemistry of Phosphorus Compounds. John Wiley and Sons, Inc., New York, pp. 617–658.

Preiss, J., 1966. GDP-mannose pyrophosphorylase from Arthrobacter. In: Neufeld, E.F., Ginsburg, V. (Eds.), Methods in Enzymology, vol. VIII. Academic Press, New York, pp. 271–275.

Rösti, J., Barton, C.J., Albrecht, S., Dupree, P., Pauly, M., Findlay, K., et al., 2007. UDP-glucose 4-epimerase isoforms UGE2 and UGE4 cooperate in providing UDP-galactose for cell wall biosynthesis and growth of *Arabidopsis thaliana*. Plant Cell. 19, 1565–1579.

Sanwal, G.G., Preiss, J., 1969. Sugar nucleotides and nucleotide-peptide complexes of *Chlorella pyrenoidosa*: isolation and characterization. Phytochemistry. 8, 707–723.

Shen, L., Preiss, J., 1964. The activation and inhibition of bacterial adenosine diphosphoglucose pyrophosphorylase. Biochem. Biophys. Res. Commun. 17, 424–429.

Weiner, H., Stitt, M., Heldt, H.W., 1987. Subcellular compartmentation of pyrophosphate and alkaline pyrophosphatase in leaves. Biochem. Biophys. Acta. 893, 13–21.

Yang, T., Bar-Peled, M., 2010. Identification of a novel UDP-sugar pyrophosphorylase with a broad substrate specificity in *Trypanosoma cruzi*. Biochem. J. 429, 533–543.

Yin, Y., Huang, J., Gu, X., Bar-Peled, M., Xu, Y., 2011. Evolution of plant nucleotide-sugar interconversion enzymes. PLoS One. 6 (11), e27995. Available from: http://dx.doi.org/10.1371/journal.pone.0027995/. (accessed 10.02.16).

Zhang, Q., Shirley, N., Lahnstein, J., Fincher, G.B., 2005. Characterization and expression patterns of UDP-D-glucuronate decarboxylase genes in barley. Plant Physiol. 138, 131–141.

Index

Note: Page numbers followed by "*f*" refer to figures.

D

E

O

P

Q